SUMMONING SPIRITS

About the Author

Konstantinos has a bachelor's degree in journalism and technical writing and is a published author of articles and short fiction. He is also the author of *Gothic Grimoire, Vampires: the Occult Truth, Speak with the Dead: Seven Methods of Spirit Communication, Nocturnicon: Calling Dark Forces and Powers,* and *Nocturnal Witchcraft: Magick After Dark.*

A Dark Neopagan, Konstantinos has been researching the occult and practicing magick for over fifteen years. He is also a trained stage mentalist who uses his skills to debunk fraudulent affectations of the supernatural. Konstantinos often lectures on paranormal topics at bookstores and colleges. He also devotes time to singing gothic rock music, and to exploring nocturnal life both in New York City and around the country.

To Write to the Author

If you wish to contact the author or would like more information about this book, please write to the author in care of Llewellyn Worldwide, and we will forward your request. Both the author and publisher appreciate hearing from you and learning of your enjoyment of this book. Llewellyn Worldwide cannot guarantee that every letter written to the author will be answered, but all will be forwarded. Please write to:

Konstantinos
℅ Llewellyn Worldwide
2143 Wooddale Drive, Dept. 978-1-56718-381-8
Woodbury, MN 55125-2989, U.S.A.
Please enclose a self-addressed stamped envelope for reply,
or $1.00 to cover costs. If outside U.S.A., enclose
international postal reply coupon.

SUMMONING SPIRITS

THE ART OF MAGICAL EVOCATION

Konstantinos

Llewellyn Publications
Woodbury, Minnesota

SECOND EDITION
Eighth printing, 2014

FIRST EDITION, six printings

Cover design by Lisa Novak
Interior illustrations by Tom Grewe and Maria Mazzara, based on drawings by Konstantinos
Book design and layout by Kimberly Nightingale
Editing by Rosemary Wallner

Cataloging-in-Publication Data
Konstantinos, 1972–
    Summoning spirits: the art of magical evocation / Konstantinos. — 1st ed.
      p.   cm. — (Llewellyn's practical magick series)
    Includes bibliographical references and index.
    ISBN 1-56718-381-6 (pbk.)
    1. Magic. 2. Evocation. 3. Spirits. 4. Incantations.
    I. Title. II. Series.
BF1611.K66   1995
133.4'3—dc20
                                            95-24724
                                              CIP
ISBN-13: 978-1-56718-381-8

Llewellyn Publications
A Division of Llewellyn Worldwide, Ltd.
2143 Wooddale Drive
Woodbury, MN 55125-2989
www.llewellyn.com
Llewellyn is a registered trademark of Llewellyn Worldwide, Ltd.

Other Books by Konstantinos

Gothic Grimoire

Nocturnal Witchcraft: Magick After Dark

Nocturnicon: Calling Dark Forces and Powers

Speak with the Dead: Seven Methods of Spirit Communication
　　　(previously titled *Speak with the Dead: 7 Methods for Afterlife*
　　　Communication and *Contact: 7 Methods for Afterlife*
　　　Communication)

Vampires: The Occult Truth

Contents

INTRODUCTION

The magician felt a surge of excitement run through him as he picked up the leather-bound book. He carefully opened the old diary to the section marked "Conjuration," and began to read by the red light of the filtered lamp on the altar.

When the oration was completed, the magician glanced at the painted wooden triangle he had positioned outside the magic circle. Toward the center of the equilateral triangle, smoke rose from a brass censer in a steady stream, filling the entire room with the scent of peppermint. Scattered about this glowing bowl were pieces of iron, garnet, and red jasper; to the right of the censer stood a metal figurine of a scorpion that cast moving shadows on the floor as the glow of the coals illuminated it.

Slowly, the magician's gaze fixed upon the small object at the base of the triangle. The red light in the room, combined with the faint glow of the censer, clearly showed the symbol drawn on the round piece of paper. It was this sigil that the magician began to focus on as he closed his eyes.

In a few moments, the magician held up his wand and slowly started opening his eyes. The name "Phalegh," which he had been repeating mentally, escaped his lips as a whisper, and he continued calling the Mars spirit out loud. With each repetition of the name, the magician opened his eyes a little more, and his voice grew in volume and resonance.

Hovering in the smoke before him, a tall, muscular man with glowing orange eyes was staring at the magician. He was dressed in red and held a long brilliant sword in his right hand. A low rumbling sound began to fill the room, and continued to grow louder as the figure standing in the triangle became clearer.

The magician pointed his wand at the spirit and greeted him. The evocation was a success, and the magician could now communicate with the spirit freely.

Magical evocation is one of the most fascinating yet misunderstood practices in the occult world. The idea of calling forth a spirit from another plane to visible appearance, and of consequently commanding it to perform some deed, has fascinated occultists since at least the beginning of written history, and most likely before.

But why the fascination? Ask anyone who has read a grimoire such as the *Goetia* or the *Necronomicon* and they'll tell you why. These books promise great power and wealth to the would-be evoker. Most of the spirits presented within their pages are described as being able to grant the magician a number of remarkable things, including the locations of hidden treasures, the admiration of others, supernatural abilities (such as teleportation, enormous strength, and even flight), and all forms of knowledge from languages to sciences, making it pretty clear why the practice of evocation has maintained its hold on the minds of magicians all over the world. What could be more exciting than reading a few lines from a book and having some supernatural being grant you anything your heart desires? All you have to do is make sure the words are pronounced correctly, right?

Wrong. The grimoires of ancient times weren't meant to teach someone how to do evocations. They were more like notebooks or magical diaries. A magician would only write in them the things he or she experimented with, or didn't have time to memorize. Because of this, these tomes of mystical knowledge are terribly incomplete and utterly useless to the uninitiated magician. The wordy conjurations found in them are only part of a systematic, magical process.

Of course, when I was younger I didn't know this. Like many others before me, I bought my copy of the *Goetia* (one of the books of the *Lesser Key of Solomon*) and decided to practice conjurations. Using a piece of chalk, I drew a rough facsimile on the floor of the magic circle shown in the book (boy, did that take hours), and got together some crude tools that I felt would do the job. Armed with all these implements, I took my book and began to conjure.

After three repetitions of five different conjurations, which took about an hour to get through, I was rewarded with little more than an intense headache from trying to read by the light of two candles. My dream of becoming a powerful magician was shattered at the age of fourteen, and it was almost a whole year before I began looking into this aspect of the occult again.

The works of Franz Bardon, the brilliant occultist, rekindled my interest in magical evocation. Bardon had a few theories on how evocations work that made a lot of sense. I took what I learned from him and began a five-year search for other theories and techniques in hopes of coming up with a method of evocation that worked. Sure enough, with a little bit of research, a lot of experimentation, and an enormous amount of initial failures, I found two distinct types of magical evocation that work remarkably well. But before identifying these two forms of evocation, it is important to establish a working definition of what evocation really is.

Evocation can be defined as the calling forth of an entity from another plane of existence to an external manifestation in either the astral or physical plane. Evoked beings are brought closer to the magician, but never within himself or herself. This is what separates evocations from invocations. In an invocation, the magician brings some foreign intelligence within himself or herself, and allows the entity to speak through his or her body. Channeling is a well-known form of invocation.

In an evocation, however, the magician brings the entity to a plane where the magician can view it and communicate with it. Evocation is therefore an external manifestation of an entity, as it occurs outside of the magician's body. This manifestation can take place in either the astral or physical plane, depending upon the type of evocation performed.

Evocation to the astral plane is when an entity is brought to the nearby astral plane, where a trained magician or clairvoyant can view it and establish contact. An excellent tool for "seeing" into the astral plane is the magic mirror, and it is usually employed in this type of evocation. This type of magical evocation is the subject of chapter 7.

Evocation to the physical plane is the more difficult of the two to master. When evoking an entity in this manner, the magician must facilitate the full materialization of the being on the physical plane. For this to be possible, the room has to be made to agree with the entity's "nature." Once this preparation is made, the magician could then bring the spirit through the planes to this one. The secrets behind this potent technique are revealed in chapter 8.

Now that we have a working definition of what magical evocation is, we should be able to illustrate what it is not rather simply. This next statement may seem a little odd, but trust me, I'll explain it: Magical evocation is not as easy or hard as the grimoires make it seem. The process of evocation entails more than just reciting some lines from a book. There is a systematic process to the art that the authors of the ancient grimoires knew, but didn't feel like sharing. In fact, not only did they not give the reader enough information to make the rituals work, they actually fabricated bizarre practices and "rituals" to throw the uninitiated off the track. While some of them were simply meant to be a waste of time, most of them were created to deter someone from ever trying an evocation in the first place. For example, the *Grimoire of Honorius* would have you prepare for an evocation with almost a month's worth of meaningless rituals, including two animal sacrifices, the preparation of a lambskin covered with dozens of incoherent symbols, and traveling to fields and "secret" places to bury various parts of the animals' corpses.

Even though all of these so-called "preparations of the operator" were absolute nonsense, they were worded in a way that made people believe in their potency, and I'm sure quite a few people did try the rituals, with no results. The truth is, magical evocation requires no animal or human sacrifice, no blood, no bathing in rivers, no burying of rooster feathers at a crossroad, and absolutely no pacts with demons. There is nothing evil or sadistic about this magical art at all. Magical evocation is a positive and beneficial experience.

This book is your guide to the art of magical evocation. It is the only book you'll ever need to learn this ancient practice, and it is unique in that it covers every aspect of magical training necessary to obtain results. Even if you've never practiced magic before, you can still safely perform evocations by first practicing the magical training exercises in the following chapters.

The names and seals of many useful spirits are found in ancient grimoires. Some of the spirits are so vaguely described, however, that a magician summoning them for the first time has little idea of what to expect. So to make things easy, in chapter 9 I've included a listing of entities and their sigils that I have personally evoked and found useful. These entities are fully explained, including their appearances, areas they are knowledgeable about, and tasks they could best perform. This way you can begin conjuring without wondering what it is you're calling, and more practically, without another visit to the bookstore.

A magician must employ several tools to successfully practice evocations. The construction, magical preparation, and use of these tools are all described in the following chapters. Once you prepare your tools and develop your magical consciousness, learning how to evoke entities is relatively easy.

The preceding paragraphs contain many magical truths that contradict what most people believe to be true. As I said earlier, there are many misconceptions about magical evocation. These include the idea that evocation is evil, that it is necromancy, that it is used to sell one's soul to the Devil, and, most interestingly, that it is easy to do (read from a book and a spirit appears).

So where did all these misconceptions come from?

A good number of occult misconceptions originated in the West Coast of the United States, or more accurately, Hollywood. Let's face it, moviemaking is a business, and as a business it has to make money.

Movies aren't supposed to be true to life, just entertaining. When people go to see a movie with a paranormal theme, they're not looking for inspiration or philosophical teachings, they're looking for entertaining horror or fantasy. In fact, the most successful horror films or novels are the ones without a shred of occult truth to them.

Let me make something clear before I go any further. I have nothing against horror or fantasy. I find them to be the most entertaining types of fiction. But that's all they are—fiction. The fact that many people get their ideas of what magic and occultism are from fiction explains why so many people are misinformed about real magic. For a movie to be entertaining, wizards have to be able to shoot lightning from their fingertips, televisions have to be able to suck children into them, and anyone can call forth a demon by reading a few funny sounding words from a crumbling book.

Of course, movies weren't always around to distort the truth behind magical evocation. As I've mentioned earlier, it was the grimoires themselves that did a lot of the distorting. Some of the "rituals" described in the ancient books of magic make Hollywood versions of evocations seem almost feasible. So let's say for the moment that movies, novels, and the misleading portions of grimoires are responsible for the idea that reading from a book will summon a spirit to visible appearance. That would take care of one of the big misconceptions about evocations.

Another misconception people often have about magical evocation is that it is evil to summon spirits. Of course, most people thought of evocation as necromancy, which is completely different. Necromancy is the calling forth of the spirits of the dead. The entities summoned in evocation are not dead, they were simply never alive in the first place (see chapter 1). Mediums are usually the ones concerned with contacting the dead, and while they sometimes appear to the medium or person being consulted, they are not evoked to physical appearance. It is the spirit's choice whether or not it wishes to appear. If you are interested in learning about this type of spirit communication, check out my book *Speak with the Dead: 7 Methods for Afterlife Communication* (Llewellyn, 2003).

Finally, we come to a belief that was very common in medieval times, and which, thanks to Hollywood, is still popular today. This misconception has its roots in one of the most famous tales of all time: the legend of Doctor Faustus. Of course, I'm talking about the idea that evocation is nothing but the conjuring of demons to help you make a pact with the Devil himself.

The tale of Faustus has been told in many different forms. It first appeared in 1587 as a German booklet entitled *Historia von D. Iohan Fausten.* In 1592 it was translated into English with a title that leaves little to the imagination of the reader: *The Historie of the damnable life, and deserved death of Doctor Iohn Faustus, Newly imprinted.* This booklet was the basis of all the Faustus books, plays, and poems, until Goethe made a big change, but we'll get to that momentarily. Christopher Marlowe's play *Doctor Faustus* is similar to the booklet and the following is a summary of Marlowe's tale.

Faustus (Faust in Goethe's version of the story) was an established Doctor of Theology at a German University. Whether or not he was a real man is still unclear, although there were a few men of the fifteenth century who fit his description. Doctor Faustus was dissatisfied with the knowledge available at the time and looked to the occult to find the truths of the universe. He is said to have studied various forms of magic, but the only apparent success he ever had was when he summoned the demon Mephistopheles (Mephisto in Goethe's version).

Mephistopheles promises Faustus all the knowledge of the universe, transportation to any place in the world, riches, and his own personal obedience to Faustus. In exchange Faustus must make a pact that after twenty-four years of life in this manner, the Devil could come and take his soul. Faustus makes this pact and comments on how he doesn't believe in Hell or damnation. Here Mephistopheles tells him, "Aye, think so still—till experience change thy mind!"

For the rest of the tale, Faustus revels in small feats of magic. He conjures spirits for the pleasure of nobility and friends, discusses metaphysics with the pope in a magical disguise, makes a castle appear to the duke, causes horns to grow from an insulting soldier's head, and flies over the world, learning all the mysteries of the universe, as promised.

When Faustus summons the spirit or "shadow" of Helen of Troy, however, it seems his fascination with his new powers gets the best of him. Upon seeing her he recites the famous lines: "Was this the face that launched a thousand ships, and burnt the topless towers of Ilium? Sweet Helen, make me immortal with a kiss." Many critics have commented that Faustus damned himself by becoming obsessed with this specter. When he has the opportunity to repent later on in the story, he finds it impossible to do so, possibly because he is in love with Helen. Whatever the reason, Faustus does not renounce the pact, and at the agreed time, the demons come and carry away the screaming soul of Faustus. When his colleagues come to his chambers later, they find his corpse terribly mangled.

This version of Faustus went a long way in enforcing the Roman Catholic Church's anti-magic laws. People were afraid of ending up like Faustus and believed that magic was the work of the Devil because of tales like this one. But before I go into some other "historical" tales of evocation, I want to first deal with Goethe's Faust, which contains some very interesting differences from previous Faustus tales.

The Faust theme in Johann Wolfgang Goethe's work is that of eternal striving. When Faust made a pact with Mephisto, he didn't agree to a certain date. The terms of the agreement were as follows: If Faust should ever stop striving to become a better person, then Mephisto would get his soul. From this comes the famous quotation *Zum hochsten Dasein immerfort zu streben*, or "To strive for the highest life with all my powers."

Aside from this inspirational theme, Goethe's *Faust* has several other differences from the Marlowe play and other previous Faustus stories. Faust does not summon Mephisto in this story. The only evocation Faust performs is the evocation of the *Erdgeist*, or "earth spirit." It is interesting to note that this evocation performed by Faust is actually very similar to a method of evocation I'll be dealing with later on. Faust meditates on the symbol of the spirit in a book and utters an impromptu conjuration, which causes the spirit to appear. Suffice it to say, many magicians simply meditate on spirit sigils and use them to open doorways to the astral plane and the realm of the entity. This is a type of evocation to the astral plane discussed in great detail later on.

Rather than being evoked, the spirit Mephisto is attracted to Faust because of a "bet" Mephisto made with God. Mephisto thinks he can tempt Faust away from becoming an advanced being, or adept of sorts. God doesn't seem to think this is possible and tells Mephisto: "A good man in his darkling aspiration remembers the right road throughout his quest."

After the pact is made, Faust leads a somewhat evil life, and for a while it seems as if Mephisto will win the bet. But Faust ends up striving to become better. He meets Gretchen, who pulls Faust away from his evil aspirations to some extent. She grounds Faust in a way that becomes instrumental to his salvation at the end of the story. Faust starts to perform good deeds, such as draining the sea away from a stretch of land so people could reside there.

While performing these generous acts, Faust reflects on how this feat of his is the best humanity ever knew. But instead of being satisfied, he thinks of how even better achievements are possible. In fact, while Mephisto's demons are preparing Faust's grave, Faust contemplates a more perfect state of being and saves himself with his last words spoken on the earth plane: "As I presage a happiness so high, I now enjoy the highest moment." The fact that Faust's last words were ones of aspiration angers Mephisto. He had lost his bet with God because Faust refused to be satisfied with the greatness of the moment. He strived for excellence even at the end. As Faust ascends into heaven, he is greeted by his lost love, Gretchen, who helps Faust rise up to higher "spheres."

By now you may be wondering why I included both versions of the Faust tale. I had a very good reason for doing this. Before practicing any form of magic, you should remember that the source of that magic is Divine and all uses of that magic should be to better oneself and make oneself closer to the Creator. While Marlowe's *Doctor Faustus* presents evocation and magic in an evil light, Goethe's Faust goes a long way in dispelling the teachings of the Church that say all occult practice has to be evil. In the Western Ceremonial Tradition, the whole point of practicing magic is to obtain conversation with your Holy Guardian Angel and eventually reunite with God. This is also the underlying theme in *Faust*, as the phrases "highest life" and "higher spheres" can only mean one thing—a union with the Godhead.

Goethe was very well read in the occult, and most people in Western Magickal circles agree that he was a true initiate. I highly recommend a careful reading of some of his works. You'll be surprised to find just how much practical occult knowledge is contained in them. One of the most important lessons of Faust, and the reason I included Helen of Troy and Gretchen in this synopsis, is a lesson every magician should learn before he or she tries evocations:

Never become obsessed with an entity!

Chapter 1 is all about entities, but I have to stress this warning here in the beginning. Just as in any other occult practice, obsession can be very dangerous. This is the reason most rituals have a grounding segment at their conclusion. Wiccans have the ceremony of the cakes and ale, the Golden Dawn has extensive banishing rituals followed by an almost Eucharistic ritual with wine, and sweat lodge members go so far as to run out of their lodge and dump cold water over their heads! All these rituals are very important, as they help to return the individual to a normal state of consciousness.

I can't stress enough how dangerous it is to perform an evocation and become so enthralled with an entity that you start to follow all of its advice, without questioning it first. A link is established between an evoked spirit and the magician, and it is best to stay in control of that link. If done properly, evocations are harmless, but just like anything else in the universe, they can be abused. Some doctors agree that a drink of alcohol a day could be beneficial in terms of reducing stress, but if you start drinking too much, the results could be quite disastrous. The same holds true for magical evocation and any other types of occult practice. Everything in life, both mystical and material, should be done with moderation.

Doctor Faustus became so obsessed with Helen and magic that it destroyed him. Faust, on the other hand, retained an interest in physical things and people (particularly Gretchen), and was saved. Make sure all of your mystical practices are followed by grounding practices. Simply going to the kitchen and making yourself a sandwich is good enough. It is not a good idea to perform a complex ritual late at night in your bedroom, and then hop into bed without performing adequate banishing rituals such as the ones in this book. Even then I recommend going for a walk or watching TV to return to normal consciousness.

As we have seen, the Faust legends are partially responsible for both causing and dispelling misconceptions about evocation. I mentioned earlier that the Church promoted the idea that all magic is evil. The Bible itself is full of inconsistencies regarding the communication of man with the spirit world. In the Book of Deuteronomy the people of Israel are told: "There shall not be found among you any one . . . that useth divination . . . , or an enchanter, or a witch, or a charmer, or a consulter with familiar spirits, or a wizard, or a necromancer" (18: 10–11). But in I Corinthians we are told: " . . . The manifestation of the Spirit is given to every man to profit withal. For to one is given by the Spirit . . . the working of miracles; to another prophecy; to another discerning of spirits [!] . . ." (12: 7–10).

I don't believe a greater contradiction could have been possible if it was planned. This discrepancy in the Bible reminds me of what happened one day in my Catholic high school. Someone in the class had just seen the movie *Witchboard*, and was talking about it with her friends. My teacher overheard the conversation and made a comment on how using a Ouija board was satanic. I was already immersed in occult study, and I found it interesting that she felt so strongly about what had become an almost mandatory part of adolescent get-togethers. So I made it a point to ask the school's priest what he thought about Ouija boards. He was in his twenties, open-minded, and fun to talk to, so I thought it would be a good idea to go and see him.

We talked for a good half hour, and he basically told me that there was nothing wrong with using Ouija boards as long as you didn't start acting on the information you received. "It's really just a game," he said, "I mean, Parker Brothers makes one." I couldn't immediately figure out why that statement struck me as being so odd, but it came to me soon enough. I got up, thanked him for our chat, and walked toward the door to leave. Propped up against the wall, hidden from my view when I had entered, was a Parker Brothers Ouija board!

Obviously the opinions of religious leaders vary when it comes to the supernatural. But since we're on the subject of misconceptions surrounding evocations, it is important to recognize why the Bible is partially responsible for the belief that all magic is necromancy. The Bible and the Church never address the evocation of angels or other beneficial beings to physical appearance, they only address the calling forth of the dead and of infernal demons. Why is that? Well, my guess is it would be pretty hard to come up with a reason why communicating with angels is wrong, and this would weaken the Church's anti-magic policy. So the only story included in the Bible relating to spirit communication is one of necromancy.

The story of the "Witch of Endor" is found in I Samuel (28: 7-25). King Saul, the leader of Israel, was very distressed to find out the Philistines were planning to attack his country. Saul turned to God for a sign, but received nothing. That's when he decided the best person to ask for advice was the deceased Samuel, but Saul had already banished all wizards and people with familiar spirits from the land. So Saul told his servants to find a woman who was a medium still living in the land. They told him such a woman lived in Endor.

King Saul disguised himself and went to see the woman at night. He asked her to call forth a spirit whose name he would give her. She protested at first, saying the king had other mediums and wizards cast out of the land, and that she would die for doing what he asked. But Saul swore no one would ever punish her for her actions.

The "witch" called forth the spirit of Samuel as Saul asked her to do. Samuel seemed angry at Saul for calling him. When Saul told him how God didn't send him a sign as to what he should do, Samuel told Saul that since God was now his enemy, he should expect no more Divine signs. Samuel went on to prophecy the death of Saul and the defeat of Israel.

I found a version of the Bible that had a footnote at the bottom of the page this story was on. The footnote said that the "witch" had no power over spirits, and that God let the spirit appear so it could talk with Saul. The author of this disclaimer wanted to make sure people did not believe in the power of magic. But the author did not think very carefully about this statement before making it. In a way, this footnote states a known occult principle. It is the power of God that lets a spirit appear, because when a magician stands in the center of a magic circle and invokes the Deity, he or she becomes God, magically speaking, and it is with the power of God that spirits are thus evoked.

Most of the religious texts in the world disagree on what occult practices are acceptable. An aspiring magician, no matter what religion, should not let this fact bother him or her. Magic, like everything else in the world, comes from God. If used for the greater good of all, magical practice is a tribute to God, as it brings one closer to the light. The Bible, like other religious texts, contains a wealth of spiritual advice. The fact that it doesn't properly address the art of magical evocation is no reason to renounce the Bible or any other inspired work. The truest adepts and masters are the ones who see the truths in every religion.

So far, we've managed to locate the origin of and dispel some of the more common misconceptions about magical evocation. We've also seen how many of the causes of the misconceptions contain valuable occult lessons. The following is the last "historical" example of evocation we'll be looking at before we move into the practical aspects of the art. It is the best known example of evocation among occult circles, and rightly so, because the result of this series of evocations was the creation of an entire magical system.

In 1582, two men formed a close association that would forever change the world of magic. The men were John Dee, a Royal Astrologer for Queen Elizabeth I, and Edward Kelly, a virtual scoundrel with impressive crystal-gazing abilities. This unlikely union was formed when Kelly went to Dee's house and demonstrated his ability at skrying.

The type of evocation the two used for the following seven years is the first type I will be teaching in this book—evocation to the astral plane. Dee, who was a well-trained magician, would evoke entities to the astral plane where Kelly could see them in his shewstone. Through these magical experiments, Dee and Kelly contacted angelic beings who taught them a magical language called Enochian.

This language was supposedly used for magical purposes in ancient times but was lost in the ages. The angels showed Kelly tablets in the shewstone, which Dee subsequently transcribed. These tablets contained the names of magical beings in the Enochian Universe that the angels pointed out, letter by letter and backward. The Enochian language is so powerful that spelling the names backward ensured one of the entities wouldn't show up uninvited. To some extent this was a good idea, because reciting a magical name of an entity could result in some form of psychic contact with that entity, and this uncontrolled type of contact is anything but desirable.

Dee and Kelly's method of evocation was very successful, and it unearthed a system of magic that is still widely used today. Gerald and Betty Schueler have written what I feel to be the finest books on Enochian Magic. The first one in the series is titled *Enochian Magic: A Practical Manual* (Llewellyn, 1990). The authors present this type of magic in usable form, and you may want to try summoning some of the Enochian entities described in these books after you've had success at practicing evocations.

Of course, there are other examples of evocation found in legends and history, but listing all of them is beyond the scope of this book. The stories in this chapter were chosen for both their accuracy and inaccuracy in portraying evocation, as I wanted to make sure both the truths and lies about the art were revealed.

Now that it's clear evocations are not as easy as the grimoires make them seem, does this mean we could trust the promises and rewards described in their pages as well?

Countless magicians throughout the ages, myself included, have learned that with a little determination and patience, the rewards of this esoterical art are nothing short of awe-inspiring. Evocation has to be one of the most powerful (if not the most powerful!) forms of magic ever practiced. Raymond Buckland said in his *Complete Book of Witchcraft* (Llewellyn, 1990 and 2002), "It [evocation] is like trying to hook up a 1,000-volt power line to run a transistor radio!" This may be true, as many simple magical goals are easily attained with the use of talismans or candles. But what happens when you want to do something a little more spectacular? Transistor radios play music well, but let's face it, they just don't cut it at a party where fifty people are talking and laughing. You need something with a little more power.

As we'll see in the following chapters, entities can help the magician in many ways. These spirits can teach you much more than what is promised in the grimoires. They can also show you how to perform personal rituals that develop magical abilities you want to obtain.

Now, before we get into the training that will facilitate the evocation of entities, it is important to understand what these beings are. There are several different types of spiritual beings in the universe, and not all of them are exactly trustworthy. Chapter 1 will illustrate what entities really are, which ones to avoid, and how to work with the beneficial ones.

chapter one

THE NATURE OF ENTITIES

Planetary intelligences are very powerful and can help the magician in a number of related tasks simultaneously.

T he word "entity" has been mentioned several times in the introduction, and rightly so, because the whole point of performing an evocation is to come into contact with one of these beings. But what are entities, really, and where do they come from when called?

Without an understanding of the nature of entities, it is incredibly difficult to perform an evocation and virtually impossible to control the outcome of one. It is for this reason that this chapter is devoted to explaining the inhabitants of the unseen world. Like many other occult truths, however, the nature of entities has been explained both correctly and incorrectly over the years. To understand why this is so, it is important to first understand who originated these theories.

Throughout the ages there have always been two kinds of occultists: armchair theorists and practicing magicians. The former have, in the past century, attempted to explain occult phenomena using the science of the time, neglecting the fact that occult science is a science of the future. As a result of the efforts of these armchair occultists, all sorts of psychological theories about magic have surfaced. These concepts are hopelessly flawed, as their creators were basically guessing about a topic they didn't understand. Practicing magicians, on the other hand, have experimented with magical techniques and achieved repeatable results. These tested theories are the ones the student of the occult will find most useful.

The theory that entities only exist in the magician's mind originated with the armchair occultists of history. According to them, evocations do nothing but bring these entities "up" from one's subconscious, and "out" into seemingly external appearance. Followers of this teaching feel all information gained by evocations is the result of some type of telepathy, and that materializations witnessed by a number of practitioners are the result of some type of telepathic projection on the part of the magician performing the ritual. To someone who has never practiced magic, this concept could seem feasible. But to a trained magician, the flaws of this theory are immediately obvious, for a number of reasons.

In philosophy, there are two concepts known as the Efficient Cause and the Final Cause. The Efficient Cause of an object is that which caused it to originate, and the Final Cause is the purpose of the object. Upon applying these principles to mystical study, a great number of occult secrets can be learned. Practicing magicians realized this, and used these ideas as a basis for their experiments in trying to identify what entities are.

The Efficient Cause of everything in the universe is God—the Infinite and Divine Providence. Of course, it can be argued that we ourselves have created magnificent things and that certain natural processes create things all around us. For example, plants and animals

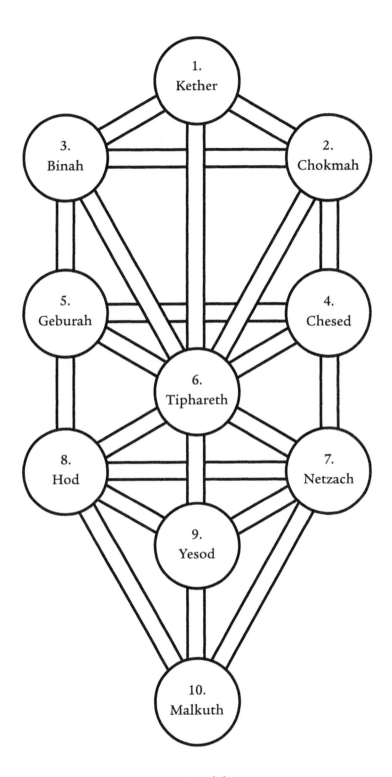

FIGURE 1.1

reproduce, tectonic plate movement creates mountains, and artists paint masterpieces. But Divine guidance affects all of these things. The inspiration for a painted masterpiece comes from God, as does the ability for plants and animals to reproduce, and mountains to form.

If it is accepted that Divine Providence is responsible for all creation, then the next logical step is to try and figure out what certain things were created for. The best way to facilitate this is to look at a Kabbalistic map of creation called the Tree of Life (figure 1.1). Each of the Ten Sephiroth, or spheres, on the Tree serves a specific purpose in maintaining cosmic order, and as a result, each are made up of different types of energies related to their tasks. For example, the energies of the lowest Sephirah, Malkuth, help maintain order in the physical plane, while Yesod governs things related to lunar energies and the astral plane.

Just as these Sephiroth were created to serve a specific purpose, the entities of the universe were also created to fulfill some type of "office." Each of the Sephiroth, for example, is "inhabited" by beings that are in agreement with the energies represented there. The angels were created to oversee certain of these aspects, and for this reason they will maintain these posts forever. While humans have the ability to advance spiritually and achieve higher and higher levels of being, the entities and angels of the universe are fixed to their appointed tasks as part of God's way of maintaining order in the cosmos. This is the Final Cause of all entities.

How could entities exist only in the mind of the magician if we accept the above as true? The answer is simple: They can't. Neither the existence of the universe nor the existence of its inhabitants depends upon the presence of one human. Whoever actually wrote the *Goetia* is no longer living, yet the beings described in the book can still be conjured. The spirits he or she worked with still exist, independent of him or her—even if this author created them in the first place.

By now it should be clear that the beings of the unseen world are not accidental creations. They were created with specific attributes to help them accomplish their assigned tasks, and exist completely independent of us. However, just as some forms of physical creation are accomplished through human mediums, the creation of entities is sometimes facilitated by trained magicians. Like a painter who takes Divine inspiration and creates a marvel on a canvas, a magician can sometimes channel Divine energy into the creation of a completely new entity. This magical construct, commonly known as an egregore, is an energy being created by the magician to carry out a specific task. Egregores, like other entities, are completely independent of the magician once created.

Whether an entity is created or conjured, the fact it is independent of us does not mean we are not responsible for its actions. If one were to command a being of any type to perform an evil act, then he or she would be responsible for the karma of that evil act. It goes without saying that this type of black magic would eventually destroy the person who practices it.

Sometimes, creating an egregore can be dangerous, regardless of the use a magician puts it to. This is because egregores grow stronger the longer they are allowed to exist. While an existing entity can be banished back to its realm of origin once its task is completed, an egregore has no place to go back to. It is for this reason that magicians usually give egregores a set amount of time to "live," and after this time runs out they are ritually destroyed. This prevents them from becoming more and more powerful in the astral plane, until they are virtually uncontrollable. The legend of the golem illustrates this possibility in an accurate, yet allegorical way, and what follows is a short summary of the tale. I still recommend, however, a very careful reading of the actual story before attempting this type of magical creation.

The story of the golem takes place in Prague, where Jewish people lived in constant fear of the Blood Accusation. This was the belief that Jews needed the blood of Christians to perform certain Passover rites. Every year around Passover, riots would break out, and false accusations would be brought up against the Jews. Rabbi Judah Loew, a village elder, felt someone might be planning the destruction of his people, so he prayed to Heaven for help.

Rabbi Loew was told in a dream to make a golem out of clay to protect his people. Following this advice, the rabbi and his pupils constructed a man-shaped clay figure. The rabbi was well versed in Kabbalistic magic and enchanted the creature to give it life. He named it Joseph.

Joseph did as he was instructed and protected the Jews from physical harm. He also investigated some of the accusations brought against the Jews so, eventually, the truth behind most of these allegations was exposed. Christian people were soon shown how the Jews had been framed all along. After the monk Thaddeus was exposed as the mastermind of the plot to destroy the Jews, Rabbi Loew's people were no longer persecuted.

Since the golem had done such a good job, Rabbi Loew and the others grew accustomed to having him around, and one day they forgot to assign him a task. Joseph wandered around the village and saw a flower in a windowsill, which he then plucked. Consequently, a woman in the house screamed at him, and Joseph became uncontrollably angry. He ran around the streets, causing all sorts of damage to the buildings and scaring the villagers.

The rabbi heard what was going on and tried to stop the golem. With a great amount of difficulty he managed to bring Joseph under control again. Along with his pupils, Rabbi Loew brought the golem to his house, where he ritually destroyed him. The clay body was then stored in a secret room, where no one could find it, to ensure Joseph would never be brought back to life.

Even though the golem was a clay creature and not a spirit, it is still an accurate metaphor of a real egregore. Once named and created, the golem was a servant to Rabbi Loew, yet when the rabbi neglected to give it instructions, the golem went off on its own and caused mischief. The warning of the story is clear. These beings have no place to return to once their purpose is accomplished, and they must not be allowed to linger. Cruel as it may seem, they have to be destroyed.

You will find a method for creating egregores later in this book. The actual process of creation contains a segment that establishes the limit of the entity's existence, and gives a method for easily terminating it when the time comes. There is nothing evil about this practice, as egregores are simply energy constructs created for one task. Upon their dissolution, the energy built up in egregores returns to the cosmos.

Types of Entities

We have seen how entities could originate in the minds of humans, and how they can only exist independent of them. While the purpose of an egregore is determined by a magician, the purpose of other entities in the universe has to be discovered. Egregores were dealt with first because they are relatively easy to understand. The nature of pre-existing entities is a little more complex, however, because there are so many different types of them to learn about.

By way of introduction, following is a list of some of the types of entities found in the universe.

Planetary

These are some of the easiest entities to work with. Each of these beings represents the magical and astrological aspects of a planet, and can use these forces to aid the magician. For example, a spirit from the Venus sphere would be helpful if a magician needed advice in matters concerning love and friendship, while a Mars intelligence would aid the magician in gaining courage and willpower. Planetary intelligences are very powerful and can help the magician in a number of related tasks simultaneously. The attributes of each of these planet's intelligences closely correspond to specific Sephiroth on the Tree of Life. Therefore, to avoid repetition, I will be exploring the powers of the inhabitants of the Sephiroth and planets in the following description of angels.

Angelic

These entities are, without a doubt, the most beneficial ones a magician can work with. They are more than willing to help a magician with a task they are proficient with, and answer truthfully any questions put before them without hesitation. Angels were assigned to certain tasks by God when they were created, and unlike us, can never advance from their current spiritual level of development. The Almighty intended for them to be keepers of universal order, and as such, they possess great power in areas they oversee. Many angels, especially the archangels, are assigned to individual Sephirah on the Tree of Life, and it is important to understand the attributes of their respective spheres before evoking them. The following is a listing of the Ten Sephiroth, containing the name of the planet each Sephirah represents, the name of each ruling Archangel and Angelic Order, and some of the general abilities of the angels and planetary intelligences that inhabit each sphere:

1. **Kether (Uranus):** The Archangel of this Sephirah is Metatron and its Angelic Order is Chayoth ha-Qadesh. Beings representing these forces can help the magician in matters concerning presently understood technology, and can also aid him or her in developing new inventions. Kether and Uranus represent intense mental energies, and as a consequence, the entities of these realms possess great intellectual abilities, which they can help a magician develop. Frequent contact with the angels of Kether and repeated use of this Sephirah's energies results in a close attunement of oneself with his or her Higher Self, which is the true goal of all magic.

2. **Chokmah (Neptune):** The Archangel of Chokmah is Raziel and the Angelic Order of the sphere is Auphanim. The spirits of this Sephirah can help the magician perfect his or her psychic abilities to an amazing degree by teaching him or her special rituals to enhance whatever faculties are desired, including telepathy, clairvoyance, psychokinesis, and even the ability to manipulate various energies. Entities from this sphere are extremely knowledgeable about the many forms of energy in the universe, and can pass this information on to the magician.

3. **Binah (Saturn):** The Archangel of this Sephirah is Tzaphqiel and the Angelic Order is Aralim. Saturn and Binah represent some of the most misunderstood energies in the universe. Destruction, death, and limitation are the forces the entities of this realm oversee, and contrary to popular misconception, they are not negative forces. After all, these forces could be used in positive ways. The beings of this sphere could "destroy" disease, "kill" bad habits, and could limit the potency of certain

influences in a magician's life. For example, the effects of probability could be affected by limiting the amount of bad luck in one's life. In other words, contests are easily won with the aid of these beings. Entities from Saturn and Binah can also teach the magician many hidden truths, from little-known occult knowledge to accounts of what occurred in ancient cultures.

4. **Chesed (Jupiter):** The Archangel Tzadqiel and the Angelic Order Chashmalim rule this Sephirah. Just as Jupiter was the king of the mythological gods and goddesses of ancient Rome and Greece, the spirits of this realm represent various "king-like" energies. A magician who works with Chesed entities could acquire great riches and material pleasures. These beings could also teach a magician how to rule wisely, whether it be in a business or a kingdom, and could guide him or her to career success and fame. One should never underestimate the power of these beings. Entities from this realm could make a magician very happy, but he or she must be careful not to abuse their powers. Greed is not looked upon kindly by Higher Powers, and one should try to ask for only what is really needed.

5. **Geburah (Mars):** The Archangel of this fifth Sephirah is Kamael and the ruling Angelic Order is Seraphim. The spirits of this emanation possess an enormous amount of energy and vitality. They can literally charge a magician with energy and strength to perform whatever tasks he or she needs to accomplish. Certain entities from this Sephirah and planet can actually show a magician how to become physically and emotionally stronger. They can also teach one how to become courageous, determined, and aggressive, with the ability to overcome any obstacle. Occult students in military or law enforcement careers should especially find Mars and Geburah intelligences helpful.

6. **Tiphareth (Sun):** The Archangel of this Sephirah is Raphael and its Angelic Order is Melekim. Beings from Tiphareth and the Sun can help a magician obtain harmony in his or her life by granting health, and reasonable amounts of money and friendship, and by showing him or her how to find peace through wisdom, all of which are essential to happiness and success. Conversations with these spirits are nothing short of inspiring, and they can teach one about several beneficial mystical secrets, including the miraculous art of healing.

7. **Netzach (Venus):** The Archangel of Netzach is Haniel and the Angelic Order of the Sephirah is Elohim. Love, pleasure, art, fertility, and friendship are the energies represented in this sphere. Netzach and Venus entities can grant all of the above to a magician who seeks them. In addition, they inspire artistic creations, and many beautiful works of literature, music, and art were created with the aid of these beings. When requesting the help of these entities in matters of love, however, the magician should be very careful. Never, under any circumstances, should a magician try to force someone to love him or her magically. This is an evil act, and these pure beings will not assist a magician who attempts it. Instead, they will help one find a mate or friend who is suitable, and will help draw that person into one's life.

8. **Hod (Mercury):** The Archangel of this eighth Sephirah is Michael and its ruling Angelic Order is Beni Elohim. The intelligences of the Mercury and Hod sphere are some of the best teachers a magician can ever hope to work with. From help in everyday scholastic subjects to unlocking the secrets of alchemy and divination, these spirits can teach their evoker the secrets of the universe.

9. **Yesod (Moon):** The Archangel of Yesod is Gabriel and the Sephirah's Angelic Order is Kerubim. This sphere represents the astral plane, and its inhabitants can teach the magician several types of astral magic. Whatever is created on the mental plane goes through the astral and eventually manifests on the physical plane. These spirits can teach the magician how to manifest objects seemingly "from thin air." Yesod and Lunar intelligences can show one how to explore the astral plane and other dimensions of reality using true astral projection. They can also teach the secret arts of dream magic and prophecy.

10. **Malkuth (Four Elements of Earth):** The Archangel of this Sephirah is Sandalphon and the ruling Angelic Order is Ashim. The powers possessed by the beings of this realm are the same as those of the elementary spirits and are explained in the next heading.

Elementary

Spirits of this type are relatively easy to contact if a magician has an understanding of the magical elements they represent. These intelligences are very specialized, and unlike planetary spirits who often possess many areas of knowledge, each elementary usually has only one area of expertise in which it is proficient. This may seem like a weakness, but actually elementaries perform assignments within their power so quickly and effectively that most magicians do not mind the trouble of looking for the right one to conjure. Following are the powers of the elementaries along with the name of each type:

1. **Earth (Gnomes):** Earth Elementaries can help the magician acquire riches, material goods, and better jobs or promotions. They are also excellent teachers and can instruct one in the esoteric properties of gems and minerals. Like all other elementaries, Gnomes can provide the magician with excellent familiar spirits that will obey him or her.

2.   **Air (Sylphs):** Sylphs are excellent teachers and initia-
tors. They can help a magician learn almost anything by
teaching him or her unique ways of studying and by
actually influencing his or her mind to make absorbing
information easier. Air intelligences can also show a magician how to telepathically contact anyone in the world.
Familiars of this type are valuable servants, as they can
show the magician how to control the winds and practice levitation.

3.   **Fire (Salamanders):** Beings of the Fire Kingdom can
show an evoker how to control the energy and vitality of
this element to bring about change in the world. From
protecting the magician to giving him or her the power
to accomplish what needs to be done, these beings are
extremely helpful. They can also teach the magician how
to manipulate actual heat and flame, which is an awe-
inspiring power.

4.   **Water (Undines):** These gentle spirits can help a magician with relationships of all kinds and can help him or
her solve all kinds of discord among loved ones. Undines
can bring great peace to a household and are some of the
friendliest spirits a magician will encounter. Unfortunately, they possess such great beauty that many magicians
become infatuated with them. As I mentioned in the
introduction, this could be very dangerous. Remember,
no matter how attractive and enticing a spirit may be, it is
still just that, a spirit. Try to limit the amount of contact
you have with a particular Undine until you have mastered control of the art of evocation. Undine familiars can
teach one how to control rain and fog and are especially
helpful to magicians who are involved in sailing, fishing,
surfing, or swimming.

Demonic

These are the hardest types of beings to control and rightly so, because they are by nature antagonistic to the magician. The last thing they want to do is help an agent of the Light perform a task. Even though they could be commanded to do so, the chances of receiving false information and weak results are increased when working with demons. My advice is to avoid working with them in the first place.

Goetic

Some people feel the spirits listed in the *Goetia* should be considered demons, but I have my reasons for disagreeing. Throughout history, the gods of one group of people would always become the demons of their conquerors. This seems to occur in the *Goetia*. One of the spirits, Astaroth, is actually a thinly disguised godform of the Mesopotamian goddess Astarte. While some of these entities do seem a little on the evil side at first glance, practice shows that many of them perform useful tasks, such as healings, and teach a great number of useful things, such as languages and sciences. They are easy to command and personal experience has shown me that many Goetic entities are far from demonic in nature.

Godforms

These are beings of incredible power, and should only be evoked under the most extreme circumstances. One of the safest ways to work with these energy sources is to invoke rather than evoke them, as an invocation or assumption of a godform only allows one to come into contact with a reasonable amount of its power and influence. Depending on his or her magical development, a magician will only be able to call in as much godform energy as he or she can control. Trying, however, to externally summon a force like that of the Egyptian gods and goddesses, for example, is difficult and could be dangerous. For this reason, the evocation of godforms will not be dealt with in this book.

Olympic

The Arbatel of Magic, a medieval grimoire, introduced these entities as the rulers of 196 Olympic provinces. The energies they rule correspond to the seven magical planets, and therefore, these beings are very similar to other planetary intelligences. Some differences between the two do exist, however, and in many ways Olympian entities are more useful than most planetary ones. This is not widely known because the most important attributes of the entities found in the Arbatel are hidden behind the allegorical descriptions of each. For example, when alchemy and the philosopher's stone are mentioned, the author of this grimoire is really talking about personal alchemy. If a magician wishes to enter the path of High Magick, then his or her goal should be perfection, or adepthood. Personal alchemy is not the changing of lead to gold, but the transformation of oneself into a more spiritual being. The assistance these beings provide in this task is very powerful and for this reason each of the Olympic spirits is described in great detail in chapter 9.

Archetypal Images

This includes all beings that were brought into existence either intentionally or unintentionally through collective thought processes. Descriptions of these two types of entities follow:

1. **Intentionally Created Archetypal Images (Group Egregores):** These entities are deliberately brought into being by magical lodges to serve as guardians or energy sources for specific magical workings. Group egregores are allowed to exist as long as the lodge of their creation exists and sometimes longer. The reason they do not become dangerous, like normal egregore, lies in the method of their creation. Unlike the latter, group egregores work with an energy current. When not actively working for a magical goal, these beings exist within this current and do not wander around the physical plane causing trouble. If a magician were to align him or herself with a magical current by joining a lodge, then he or she would be able to evoke and work with these beings. A solitary magician, however, has to work

with either egregores of his or her creation or with archetypal images of the following type.

2. **Unintentionally Created Archetypal Images:** Over the years, a number of grimoires and books on magic have surfaced that some people consider to be "questionable." In other words, their origin and authenticity are unclear, making them possible forgeries thought up by an armchair occultist with nothing better to do. This doesn't mean the beings described in the pages of these books don't exist, however. The fact that people read these books and tried to contact the beings within them makes these entities real. Unintentionally created archetypal images are therefore the result of these "fake" books of magic. They are created collectively just like group egregores, except the process is involuntary and the resulting beings can be contacted by anyone. Therefore, any grimoire that's been around for a few years will contain beings that can be contacted. Even though they might not come from whatever sphere or dimension described in their respective grimoires, these spirits will still have whatever abilities they are described as having, just like egregores will have whatever attributes one gives them. As collective creations, they exist in the collective unconscious that makes up the mental plane. This is where they come from when called, and this is where they return to when banished.

Astral Beings

I am listing this type of entity last to present a warning. Astral beings are deceptively clever entities that have access to something called the Akashic Hall of Records. This is a realm or "place" where all the knowledge in the universe is accessible. Occasionally, instead of contacting the entity you are trying to summon, you may end up contacting an astral being that could pretend to be the one you are looking for. They could answer several questions put to them by tapping into the Hall of Records, and could easily fool the unprepared magician. Later in this chapter, we'll be dealing with ways of making sure of an evoked entity's identity.

Most of the mentioned entities exist in hierarchical groups. These are groups of spirits in realms or spheres ruled by powerful king or queen spirits, with less powerful entities filling in various subservient offices. Examples of these positions are found in many grimoires, which identify princes, dukes, and earls as some of these beings' titles. It is important for the magician to understand this hierarchical nature of entities before practicing evocations, as one must learn what forces control a spirit before he or she can hope to control it as well.

Whenever a spirit is described as belonging to a kingdom or as serving some kind of superior entity, the magician might not always be able to contact it with ease. If this occurs, it is likely the spirit is being detained by its ruler. For this reason, many grimoires give conjurations that command kings and queens of the spirit realm to allow their servants to appear. Goetic intelligences and elementaries are often controlled by their rulers in the above mentioned manner. Angels and planetary intelligences, however, are usually free to come when evoked, and a magician rarely has to command these beings' superiors to let them appear.

Choosing a Specific Entity

Once you pick the type of entity you want to work with, depending upon your magical goal, the next step is to pick a specific entity of that type to evoke. If your chosen entity is described as possessing whatever talent or area of knowledge you are looking for, then the evocation should be a success.

Sometimes, as I mentioned earlier, a magician can fail in trying to contact the entity he or she is trying to evoke. This can be attributed to the influence of the spirit's hierarchical superiors or to lack of concentration and preparation on the part of the magician. The closest unseen realm to us is the astral plane and, on occasion, an astral being could answer a magician's call before it can reach the distant entity he or she is trying to summon. How do you know if you are working with the right entity?

Spirits could lie using information they acquire through the Akasha and the mental plane. For this reason, a magician needs to have an accurate way of testing the identity of what he or she summons. Luckily, there is an effective system for determining an entity's identity. The first step of this process is for a magician to learn as much about an entity as possible before evoking it. Careful notes should be made of what type of entity it is, its office, the names of its superiors, its appearance, its talents, its areas of expertise, etc. Then when the entity appears, the magician should ask it what its name is. Never say: "Are you the angel Raphael?" This is practically asking an astral being to lie. Instead, ask what its name is and see what it says. If it answers correctly, then it might be what it says it is. If it gives you a name you never heard of, chances are it's not.

Sometimes the entity will answer in a kind of code. If you make an accurate list of the spirit's known attributes, then even coded answers could make sense. For example, if it says, "I am the Divine Physician," then you will know it is in fact the angel Raphael.

But no matter what the spirit says, it is a good idea to put its identity to the ultimate test. Simply ask it to sign its name, either in a book you provide, in the air, or on the surface of a magic mirror, if that is the type of evocation you're performing (for an explanation of spirit signatures, see chapter 6). For some reason, entities cannot sign any other name besides their own. This is the most foolproof of all identification methods.

In chapter 2, we'll be looking at magical training techniques designed to prepare oneself for the practice of evocation. Each of these exercises should take only a few minutes to perform and, if done properly, ensure one's success in this magical art.

chapter two

MAGICAL TRAINING

Imagine that the darkness before you is like a black fluid and that it is moving inward in a spiral fashion. Try to see this vortex forming before you.

Before attempting evocations or any other type of magical practice, one's mental faculties must be conditioned and trained. After all, the human mind is the most powerful implement in any occult practice, and once the magician masters its abilities, he or she can perform any magical task with ease. This chapter contains exercises designed to help develop magical senses and abilities in an individual. You'll find them in the order in which they should be performed.

What magical senses have to be developed? The most important ones are the abilities to "see" and "hear" astrally. Mastering these faculties ensures success in evoking spirits and communicating with them. The importance of their development cannot be overstated. While it's true that obtaining these abilities is easier for some than others, it's not an impossible process. Anyone could train his or her astral senses through steady practice and application of the following training methods.

One of the most discouraging things about trying to develop magical abilities is the lack of immediate observable results. Many occult training systems present this problem to the novice and, as a result, many students give up. I took this into consideration when I put together the following exercises. Instead of showing how to train the magical senses with repetitive and tiresome exercises, you will find how to develop these senses through the practice of other magical techniques, which will grant the training magician observable results. For example, exercise set four is a step-by-step explanation of how to skry or crystal gaze. The reader will know right away if he or she is obtaining results and, at the same time, will learn an additional valuable magical technique.

Both astral senses (sight and hearing) should be developed at the same time. For this reason I have grouped the training exercises into sets that should be performed together. I urge you to keep a magical diary of your progress. Sometimes, certain astrological influences or the phases of the moon can affect your workings. If you have a diary, you can go back and identify these patterns. Also, since an accurate recording of every evocation should be made to ensure there is no loss of valuable information, keeping a diary at this point will help develop this important habit.

Exercise Set One

The first step in training one's magical senses involves developing concentration. Nothing ruins an evocation's chances of working more than if a magician thinks about what he or she wants for dinner while reading a conjuration. Once a magician can concentrate on one idea for several minutes, then he or she can easily master magical techniques. Following is the process for developing concentration.

First, find a period of time that you can devote to your exercises every day. The best times are either before going to sleep or right after waking in the morning, as you are in a calm state of mind in both instances. But no matter what time you practice, try to make it the same time each day, because setting aside the same time helps establish a beneficial routine in your subconscious. In other words, your mind will expect a period of magical training every day at your selected time, just like it expects to fall asleep when your head hits the pillow at night.

Once you find a time to train, the next step is to find a place that is free of outside distractions. A bedroom is fine, as long as you turn off the phone's ringer, and make sure no one will knock at your door. Find a good comfortable chair to sit in and face toward the part of your room that is the least cluttered or disorderly looking. Never try these exercises in your bed! As I already mentioned, your mind expects to fall asleep when you are in bed and meditation on a mattress is a little too relaxing.

When you are ready to work, turn off any electric lights and either work in darkness or by the light of a candle or oil lamp, either of which should be behind you if used. This will create a soft glow in the room that is very relaxing. It is also helpful to burn a little incense, which helps put one into a slightly altered state of mind. Any scent will do at this point. I like using a combination of frankincense and myrrh, because the scent sets a temple-like atmosphere.

Sit in your chair and try to get as comfortable as you can, while at the same time making sure your back is kept straight. I recommend sitting with your feet pointing straight ahead and about six inches apart, with your hands resting lightly on your lap, palms down. Let your shoulders relax, but try to keep your head up straight. The slight motion of your head tipping forward should serve as a warning that you are becoming either sleepy or restless. If either occurs, it's probably best to continue at another time. Make sure you have a pen or pencil and your magical notebook handy before you go any further, so you can record your results when you are finished with the exercise.

Once you are sitting comfortably, close your eyes and begin the following rhythmic breathing exercise: Inhale deeply through your nose for a slow count of four, and when your lungs are filled, hold this breath for the same count of four. Next, exhale slowly for a count of four, and when your lungs are completely empty, keep them this way for another count of four. Then repeat the process. After a while of breathing in this rhythm, you won't have to count any longer, as your body will grow accustomed to keeping the rhythm. With practice, you should only have to go through the conscious counting for three or four cycles before your body takes over. (If you have heart problems, just breath rhythmically without holding your breath.)

The next step of this initial relaxation process is to become aware of a glowing sphere in the air before you. Don't worry about being able to visualize it clearly with your eyes closed, just know that it is there, and do your best to imagine it in a warm, golden color. When you feel ready, imagine that your body is hollow and that you are clear like glass. On your next rhythmical inhalation, pull some of the light of the sphere toward your feet for a count of four, and feel its warmth filling you. Then hold this fluid-like light within your feet for a count of four, and as you exhale, feel it glowing brighter within you. On your next inhalation, pull some more fluid light into yourself to rise and fill another six to eight inches of your legs. Hold as before, and feel this light glow brighter as you exhale.

Repeat this process until the light moves up your entire body, filling it with this warm, golden fluid. By now, your breathing should become an unconscious process, and you should be able to concentrate on the next step without counting. Remember, your breathing only has to fit an approximate rhythm, so don't worry if you are off by a beat or two.

For the first week or so that you are working at this set of exercises, perform the following step. After your relaxation is completed, imagine you are sitting on top of a cliff overlooking a river. There is a clear night sky above you. Again, don't worry if you can't visualize this clearly yet. Just know that these are your surroundings. Now look at the river as it flows by you and listen for the sound of running water. As you do this, you will probably find several distracting thoughts are trying to occupy your consciousness. Don't fight this, but instead, let these thoughts come, and try to see each of them as a positive visual symbol. For example, if your next car payment is worrying you, and you think of it at this time, see it manifest as an envelope full of money and let it drift down the river. This way you don't become too involved in each thought. Once you cancel negative thoughts in this manner, and let them drift down the river, you can then let the next distraction come to you so you can deal with it in the same manner. Never dwell on a distraction, just make it positive and let it drift away. In time, fewer and fewer distracting thoughts will invade your concentration.

Stick to the above exercise for the first week or so. When you see a significant drop in the number of distractions you experience while training, move on to the next step. Once you are seated and have performed your relaxation, imagine you are on that cliff again, but this time let an idea come to you, and concentrate on it. Give it a positive symbol, or even better, make sure it is a positive idea. Visualize this symbol, and try not to let any other thoughts enter your mind. If you have a hard time, you might want to try letting the first few ideas that come to you drift down the river as before, and then concentrate on one. Try and hold this one thought in your mind for as long as possible. You might find it helpful to hold a moving image in your mind, like a slow-turning propeller, as concentrating on the motion will help you remain focused. Each day you try this exercise, you should be able to go a little longer without any distracting thoughts.

Once you feel you can concentrate on one idea for five or six minutes, without interruption, you are ready for the next set of exercises. Remember to work at whatever pace feels most comfortable for you, and success is ensured.

Exercise Set Two

After you have achieved some success at keeping one image in your mind for several minutes, the next step is to improve your ability at viewing that image. You should also start developing your ability to "hear" sounds that accompany that image. The following exercises are designed to do both of the above, but don't worry, these are the last ones in this book that will actually seem like exercises. In set three you will start learning practical magical techniques, while developing your magical senses.

To begin, select some object you feel is simple in appearance. A pencil, match, colorful pebble, or a key all work fine. Sit in your chair at your normal training time with your object in your hand. Also, for the following exercise, make sure you have a light source in your room that you can easily adjust. An oil lamp is perfect, as you can control its brightness by turning a knob.

When you are physically prepared to begin the exercise, lower the flame of your lamp until the room is dimly lit and perform the relaxation exercises of set one. This time, however, do not try to visualize yourself on top of a cliff. Instead, open your eyes, increase the flame of your lamp, and gaze at your chosen object. Try to notice all the physical characteristics that make it unique, including its color, texture, and three-dimensional shape. Spend at least a minute familiarizing yourself with the object's appearance in this manner.

Then, when you feel ready, lower the flame of your lamp once again and close your eyes. Imagine two connected, dark curtains are floating before you. Know that beyond these curtains lies the astral plane. Will them to separate, and see them drawn apart, one to the left and the other to the right. Behind them, the color of your field of "vision" should seem a little lighter. At this point, try to visualize the object you concentrated upon earlier. Don't worry if you can't see it in vibrant color or clarity. Just know the object is there, and that you are simply trying to focus on it.

If you have been doing the concentration exercises, the image you create should not fade in and out. In the event that it does, try to keep changing your point of view in respect to the object. I hinted earlier that moving images helps to keep them in focus. This is a little-known, yet very important occult secret. By constantly rotating, enlarging, and reducing your visualizations, you ensure that they will stay in focus. At this stage in your development, moving your visualizations is one of the keys to acquiring clearer visualization ability.

After spending about five or six minutes trying to maintain an image of your chosen object, imagine the curtains closing before you, and immediately perform the following exercise.

Note: This exercise should be performed before you sit to do your daily training. Always do the visualization exercise too.

Select a sound that you find pleasing, yet uncomplicated. It must be a sound that has a portable source, so you can listen to it repeatedly. I strongly recommend you do not use a tape recorded sound, however, as the hiss and background noise on tapes become as loud as thunder when you are meditating. Either a bell with a soft chime or a soothing drum (for those interested in shamanism) is perfect for this exercise. In the description of this exercise, I will be using the example of a bell, but feel free to use whatever you feel comfortable with.

After you have done your relaxation and visualization exercises, take your bell and ring it softly. Let the bell vibrate if possible and try to feel the tingling in your ears that is caused by each chime. With some concentration, you should begin to notice that it almost feels like the air pressure changes around your ears when you hear the noise. This sensation is very subtle but it should feel like a combination of tingling and etheric pressure in your ears.

Spend a minute or so trying to isolate this feeling. When you feel you are starting to sense it, imagine that there are two dark spheres floating by your head, one near each ear. Try to bring them a little closer, and at the same time, notice the increase in etheric pressure around your ears. When these spheres actually touch you, your ears will seem to be almost physically pulsing. As soon as you feel this, you can proceed to the next step.

Ring your bell again and notice the disturbance it creates in the air pressure around your ears now. Know that these spheres are actually openings to the astral plane, and that the disturbance the bell's chime makes in them is the astral sound of the bell itself.

Now comes the important part of the exercise. After a few more rings of the bell, try to recreate the feeling in your ears of its vibrations. Each time you recreate this feeling, try to simultaneously hear the bell ring in your mind. What you are trying to do at this point is tune into the chime's astral equivalent so you can "hear" it. With practice, you should find this to be relatively easy. When you are ready to finish this exercise, simply imagine the dark spheres moving away from your head and fading away.

The exercises in this set should be practiced until you can clearly see and hear the objects of your choice in your mind for about five or six minutes. You can then move on to the following set with guaranteed success.

Exercise Set Three

Now that you have begun to develop your astral senses, the next step is to learn how to use them together. This is the purpose of this set of techniques.

No physical preparations are necessary for the following exercise. Prepare your room as before and perform your relaxation exercises. Now imagine the astral curtains are once again before you, but this time also sense the presence of the dark spheres near each ear. Once you feel their presence, the curtains will slowly open before you and, at the same time, feel the spheres getting closer to your ears. Make sure you see your field of astral vision opening while you feel the change in pressure around your ears. Achieving this simultaneous sense is important, as you don't want your concentration to fluctuate between the two astral senses.

Once you feel ready to use both of your senses together, try and visualize an object that makes a noise, such as a grandfather clock. A grandfather clock is a good choice because you can focus in on the pendulum's swing and hear it tick at a comfortable pace. As you observe the clock, make sure you time the ticks you hear with each slow swing. Also, make sure you feel the ticks resound in your ears, without any loss of visual clarity. If you've been doing the previous exercises, you'll be amazed at how easy this one is to master.

If you feel ambitious, you might want to visualize the time as being twelve o'clock and then try to hear twelve consecutive gongs. If you can imagine this, and then immediately hear the ticks resume once the last gong ends, you've mastered this exercise. Eventually, you will find you no longer need to imagine the curtains or spheres to activate your senses. Simply willing their activation is all that will be necessary. As always, keep an accurate record of your progress in your magical diary.

The following technique should be practiced once a week from this point on. You don't have to master the previous technique to

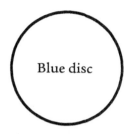

PRITHIVI (EARTH)
Divine Name: Adonai ha-Aretz
(ah-doe-nye ha-ah-retz)
Archangel: Auriel (ohr-ee-el)
Angel: Phorlakh (phor-lahk)

VAYU (AIR)
Divine Name: Shaddai El Chai
(shah-dai el chai)
Archangel: Raphael (rah-fay-el)
Angel: Chassan (chah-sahn)

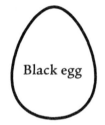

AKASHA (SPIRIT)
Divine Names:
Eheieh (eh-hey-yay)
Agla (ah-gah-lah)
Yeheshuah (yeh-hay-shoe-ah)
Eth (eth)

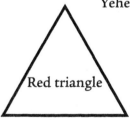

TEJAS (FIRE)
Divine Name: Yhvh Tzabaoth
(yode-heh-vahv-heh tza-bah-oth)
Archangel: Michael (mee-chai-el)
Angel: Aral (ah-rahl)

APAS (WATER)
Divine Name: Elohim Tzabaoth
(el-oh-heem tza-bah-oth)
Archangel: Gabriel (gah-bree-el)
Angel: Taliahad (tah-lee-ah-hahd)

FIGURE 2.1

start using this one, but I recommend you perform it at least three or four times before trying the following technique. Use of this next exercise leads to eventual mastery of the technique of astral traveling, which we'll deal with in exercise set five. But before we go any further, I think a short introduction to its origin is necessary.

The Hermetic Order of the Golden Dawn, founded in 1887, was probably the most influential and famous group in Western Occultism. Today several magical groups use esoteric techniques the Golden Dawn once employed, and several Golden Dawn Temples thrive across the world. Many of the magical techniques given in this book are aligned with the Golden Dawn current, and some of them are actual rituals the original Order used.

In regards to developing magical faculties, Golden Dawn initiates practiced skrying of the Tattwas. I find this practice to be a relatively easy one, and for this reason I am including it as part of the third set of training exercises.

Each of the five Tattwas represents the current of a specific magical element. Figure 2.1 shows symbols of the Tattwas and their names of power. Akasha represents the Ether or Spirit, Prithivi represents Earth, Vayu represents Air, Tejas represents Fire, and Apas represents Water. Hebrew names of power are associated with the four elemental Tattwas, including a ruling god name, archangel, and angel. Akasha has four divine names associated with it, but no angelic ones. Learning the corresponding names of each Tattwa is important before trying to skry.

Before working with the Tattwas on an astral level, you have to first make a physical set of them. This is relatively easy. Copy the shapes in figure 2.1, making each Tattwa about four inches long. Cut the shapes out of correctly colored paper and paste each one to its own piece of square, white cardboard using a small amount of rubber cement. Once the rubber cement dries, write the corresponding names of power on the back of each Tattwa. Your set is now ready for use.

To perform this skrying exercise and many of the rituals in this book, you have to first learn a simple technique for vibrating names of power. To do this, take a deep breath and slowly pronounce the word of power you are using with your entire exhaled breath. There should be no breaks in your pronunciation; the word of power should become one long, flowing, monotonous sound. If done properly, it should sound like you are almost humming the word in a single, drawn-out note. While doing this, try to "hear" this word vibrating throughout the entire universe. Practice this technique before starting the next exercise.

For this exercise, try to have an adequate yet soothing amount of light in the room. You might want to light a few extra candles or turn your oil lamp up a little. If possible, try to avoid electrical lighting. Sit in your chair with your chosen Tattwa in one hand and a blank white card of the same size in another hand, facing the correct direction of the Tattwa's element. Face north for Earth (Prithivi), east for Air (Vayu), south for Fire (Tejas), and west for Water (Apas). When working with the element of Spirit (Akasha), you can face any direction, although I recommend east as this is the direction you will be facing when doing most magical rituals. For your chosen Tattwa, make sure you know the correct words of power and their pronunciation (see parentheses after each word of power in figure 2.1).

Perform your relaxation techniques and open your eyes. Look at the Tattwa in your hand (for illustrative purposes, I'll be using the example of the Vayu Tattwa). This should not be a forced gaze, but a calm viewing of your card. When you have looked at the Tattwa for about a minute, immediately look at the blank white card in your other hand. You should see a complementary colored image of the Tattwa upon its surface. With the Vayu Tattwa, you'll see an orange disc (the reverse image of the blue disc) visible on the card. Once you see this clearly, close your eyes and try to visualize the reverse image of the Tattwa in front of you.

The next step of the process is a little tricky. Imagine that the image before you is growing to the size of a door. Once you visualize this, feel yourself moving through this symbol in a spirit body. Don't worry about how real it feels; on some level, if your visualizations are focused, you are actually traveling in the spirit vision. After you feel you have passed through this symbol, you should be able to turn around in your imagination and "see" the orange disc standing behind you. When you are ready to return, this is your doorway back to waking consciousness.

Now, having entered this elemental world, vibrate with your physical voice each name of power three times, remembering that these words resound on all astral levels as well. With the Vayu Tattwa, vibrate Shaddai El Chai, Raphael, and Chassan three times each in order (again, check pronunciations in figure 2.1). Then try to see your new surroundings. With the Vayu Tattwa, various images having to do with the magical element of air will greet you. If you sense another being is near you, do not be afraid. Simply vibrate the god name of the Tattwa, in this case Shaddai El Chai, and it can only come near you if it is a beneficial entity. An elementary guide may appear. If you vibrate the appropriate god name and it converses with you, you can trust it will take you on a very interesting tour of its elementary realm.

If no guide appears, simply view the realm and try to contemplate the rich, visual symbols before you. Try to also listen for sounds or voices in the region. Do not go wandering around without a guide, as these regions do not always follow logical, physical laws. You could end up disoriented if you try to explore a region like this on your own and could emerge from it feeling dizzy.

When you feel you've seen enough, try to leave the region through the Tattwa doorway through which you came. Do not simply open your eyes, as this too can make you feel dizzy and not altogether "with it." You should try to establish the separation between the physical and unseen world by exiting the latter the same way you entered it.

I recommend you perform this exercise only once a week and that you visit different realms each week. The elemental energies at work in the Tattwas are very strong, and you do not want to risk losing the balance of the elements in your body. For example, too much Fire could increase your temper and too much Earth could make you feel extremely lazy. Try a new element each week, keeping accurate accounts of each visit in your diary. You will find it fascinating to note the differences between your visits to each realm after five week's time.

As I mentioned earlier, you can keep practicing this last exercise even as you move on to the other exercises in this book. I recommend, however, that you master the first exercise in this set—the one that combines your two astral senses—before moving on to set four.

Exercise Set Four

The technique we will be dealing with in this set is one of the most famous ones in the occult world. Skrying, or crystal gazing, has been used by magicians and prophets for thousands of years to make predictions and communicate with spirits. The reason crystal balls and magic mirrors allow one to perform these magical tasks is simple: They open doorways to the astral plane.

When you skry, your astral sight and often your hearing attune to that plane's vibrations and allow you to perceive what is occurring there. When skrying to determine what will happen in the future or what the answer to a certain question is, the seer often views symbolic images that answer his or her question or scenes of actual incidents that have occurred or will occur. This is because when doing this type of work, we too, like the astral spirits I talked about in chapter 1, access the Akasha, where all the information in the universe is available. This is one of the beneficial uses of skrying.

Skrying can also teach you, without you realizing it, how to use your astral senses with your eyes open. This is an important ability that must be developed before trying evocation to the physical plane. It will be dealt with later on, as success in skrying is a prerequisite to effective open-eyed visualization.

To begin the practice of skrying, you have to first obtain something to skry with. A clear crystal ball, with no imperfections or cloudy areas, about four inches in diameter will work fine. This crystal can later be positioned (see chapter 4) in a way that makes it useful in performing evocations to the astral plane. Make sure it is purified before use (again, see chapter 4). For now, a crystal ball can be used by simply putting it on a black crystal stand (available where crystals are sold) and resting it on a table covered with black cloth.

Another skrying tool you can use—the one I recommend—is the magic mirror. I prefer working with magic mirrors for a couple of reasons. First, they are less expensive to make than a crystal is to buy (a clear four-inch crystal can cost eighty to one hundred dollars, while a magic mirror costs about ten to fifteen dollars to make), and, second, you can make a magic mirror as big or little as you want. To keep matters simple, the construction of all magical tools, including magic mirrors, is explained in chapter 4. If you choose to work with a magic mirror, please turn to that section to construct one.

I already stated how to set up a crystal ball for skrying. If you are using a magic mirror, however, you can simply lean it up against something on a table, use a plate display stand, or make a simple stand yourself like the one explained in chapter 4 (the instructions immediately follow the instructions for making the mirror itself). Make sure the painted side of the mirror is facing away from you, and that you are looking into the glass side.

When sitting down to skry, you must use the following types of lighting, depending upon your chosen instrument. For a crystal ball, either a candle or oil lamp will do. Just make sure the light source is behind you and that none of its glare reflects in the crystal. Use only enough light so the crystal is visible. When working with a magic mirror, you can use two candles, one on your left side and one on your right. Because a mirror is flat, it will not reflect any light from the candles as long as you move the candles far enough apart. Once this is achieved, the mirror will look as if it is almost shining. Again, make sure no reflections or lights are visible in either instrument. It is a good idea to clear your area of objects that might cast a reflection into your crystal or mirror.

After the previous preparations, you are ready to skry. Sit in your chair at your normal time, and have before you a table of comfortable height with your instrument on it. Perform your relaxation techniques and open your eyes. Gaze at the crystal or mirror as you did with the Tattwas. Again, do not forcefully stare at the instrument, just look at it in a relaxed fashion for about ten minutes. The first time you try to skry, chances are you will not see anything. Don't let this discourage you. Very few people are able to achieve a vision during their first attempt. Practicing every day, however, guarantees results within a week's time. It is okay to practice a little longer each day, but I don't recommend extending your sessions to an hour.

Early success in skrying usually takes the following form. The surface of your instrument will look like it is becoming cloudy, followed by the appearance of tiny points of light. The surprise of this occurrence might startle you the first time, and you might lose the vision. Don't worry if this happens, just try again. If you don't lose the vision, the clouds may clear up and the lights may turn into images. Let these images come to you, don't will their appearance. They might come as symbols you have to interpret, or as animated sequences you may or may not recognize. Record all these visions in your magical diary! They could prove important.

After a while of practicing skrying, you will be able to ask questions and have them answered in one of the above forms. This will enhance your astral senses enormously. For now, keep practicing skrying every day, and your work with the Tattwas once a week. When you are able to clearly view scenes in your crystal or magic mirror, you are ready for the next set of exercises.

Exercise Set Five

If you have worked the previous techniques this far, you have already made significant magical changes within yourself. This set of exercises is designed to fully prepare you for the magical work in this book. The first exercise in this set deals with the open-eyed use of astral senses. Since the rituals in chapter 3 actually help develop this ability, I recommend you start performing them daily, along with the exercises in this set. You should only use the second exercise in this set, which teaches the technique of astral traveling, in conjunction with the rituals in chapter 3.

Once you acquire the ability to use your astral senses to some degree with your eyes open, you can move on through the book and successfully perform evocations to the astral plane. Evocation to the physical plane, however, requires extremely well-developed astral senses and the ability to travel mentally to other spheres of existence. Therefore, even after you successfully perform your first astral evocation, keep practicing the following techniques to develop the abilities necessary to evoke to the physical plane. This should remain your goal, as certain magical results can only be attained through evocations to the physical plane.

To use your astral senses with your eyes open, it is best to first work in a completely dark room. Sit in your chair, perform your relaxation techniques, and open your eyes. Try to remain relaxed and just stare into the darkness before you. Imagine that the darkness before you is like a black fluid and that it is moving inward in a spiral fashion. Try

to see this vortex forming before you. Again, the motion should be easy
to visualize. If you sense any colors, try to make them out; but if you
just see dark motion, that is fine too.

After a minute of viewing this vortex, imagine it is becoming
larger and that it is moving so vigorously you can hear the dark liquid
moving. Let whatever sound you feel represents this motion come to
you. I find the sound reminds me of a roaring ocean. Spend the next
couple of minutes watching and listening.

After you feel the presence of this vortex in the above manner,
imagine that it has a center which is becoming gradually lighter than
the surrounding darkness. If you concentrate upon this, you will find
it is slowly opening. Stare into this imaginary focal point and, within
a few minutes, if you have performed the above properly, you will be
greeted by visions, almost as if you are skrying. Once you achieve these
visions, you can change them to suit your will. This is unlike skrying
in that you can force specific visions to come to you. Some clairvoyant
scenes may appear, and you may also catch glimpses of entities on the
astral plane that are curious enough about what you are doing to be
in your immediate area. Just ignore them and concentrate on your
new vision.

In this vision-inducing state, you will be able to practice forming
images with your eyes open. In addition to viewing them, you should
be able to make out any noises associated with them. If you found the
grandfather clock exercise useful, you may want to try this same procedure with your eyes open. When you are able to clearly view and
hear objects of your own creation with your eyes open, you will have
come a long way in your development and should find the rest of the
training process easy.

The following practice of astral traveling should be attempted
after each session of the above exercise. As I mentioned earlier, successfully mastering astral traveling is absolutely necessary if you want
to practice evocations to the physical plane. Having mastered all the
techniques so far, you are ready for astral evocations. I suggest, after
reading the entire book, that you begin practicing them while you are
working at the following technique. It is much easier to evoke to the
physical plane if you have already had success doing so to the astral.
We will be dealing with why this is so later on.

Before attempting astral traveling, read chapter 3 and at least familiarize yourself with the Lesser Banishing Ritual of the Pentagram. From this point on in your training, this ritual should be done immediately following your relaxation exercises at the beginning of each session and before any other techniques or exercises are attempted.

To travel astrally, perform your relaxations and then do the Lesser Banishing Ritual of the Pentagram. Sit down again and practice the first exercise in this set. When you are finished, close your eyes and imagine that you are clear like glass once again, just like in your relaxation exercise.

Begin doing your rhythmic breathing again, only this time, instead of becoming aware of a golden ball hovering in the air in front of you, become aware of a Divine white light source directly above your head. With each inhalation, pull some of this white light down into your head, keep it there while you hold your breath, and finally, imagine it becoming brighter as you exhale. Keep pulling more light down into yourself in this way until you are filled with its brilliant glow. When you accomplish this, you have formulated a body of light that you can astrally travel in.

The next step is to feel this astral body of light expand with each inhalation and contract with each exhalation. This sensation, once achieved, will help you become aware of your body of light in a new way. It will begin to take on a physical, fluid-like existence. By now you should begin to see the effectiveness of keeping mental constructs in motion. By performing this part of the exercise, your body of light will seemingly solidify. When you actually feel as if you are surrounded by a second skin, you can cease the contracting and expanding of your astral body of light and proceed to the next step.

Keeping your eyes closed, try to let your body remain rigid on its own. You have to successfully imagine that your consciousness is no longer controlling your physical body. Instead, any attempt at moving will result in your astral body of light moving. Astrally look at your body of light and try to move one of the arms free from its physical counterpart. When you see this happen, move it up in front of your face and try to move the fingers of your glowing, white hand. If you can move them and actually feel the movement, then your transfer of consciousness is complete.

Try to rise, in your astral body of light, from your chair. Will yourself to "see" your surroundings from your new vantage point. I suggest you avoid looking down at your physical body for a while, however, as the shock of seeing it may force you back into waking consciousness. Just try to notice how your surroundings look in purely astral vision. If you can move around the room and view it from different points, then you can move on to the next step. Don't worry if it takes a few sittings to accomplish this. This is the hardest part of the process.

Try to rise up through the roof of your room and out into the air. Keep rising as high as you can. If you start to feel fatigue, then simply lower yourself back to your physical body. Once you've achieved this type of success at astral traveling, you should have no doubt in your mind about being able to do it again the next day. If you wish, travel to the astral counterpart of some physical location you like to visit. Observe what is occurring there. Or perhaps you would like to visit a friend and see if he or she detects your presence. There are numerous experiments you can try once you learn how to travel in this fashion.

After obtaining success in the above steps, spend a few weeks traveling around the astral counterpart to the physical plane (see chapter 6 for an explanation of the nature of planes). When you feel comfortable in your astral form, you can try Tattwa skrying in the following manner: After leaving your physical body, once again imagine the reverse image of your chosen Tattwa in front of you. Fly with your astral body of light through the image, vibrate the names of power using your astral voice, and observe your surroundings. Your experiences in this sphere will now become much more vivid than they ever were before, and your surroundings will appear completely three-dimensional. Once you can visit an elementary sphere in this fashion, you are ready to evoke elementary entities to the physical plane.

The next step of this process will be to travel to other spheres of existence, so you can physically evoke beings from them as well. We will be dealing with this technique in chapter 8, along with the practice of evocation to the physical plane.

chapter three

OPENING RITUALS

Reach up with your right hand (or dagger) and pull this white light down to your forehead.

All magical workings, evocations included, should be performed in a consecrated circle, free of any influences that are not in agreement with the ceremony itself. To accomplish this, a magician must be able to perform banishing rituals that purify the area to be used. In this chapter, we will be dealing with some of the most powerful banishing rituals ever created: those used by the Hermetic Order of the Golden Dawn. In addition to preparing an area for magic, these rituals have another important function. By practicing them, you can learn how to raise and work with magical energy, which is a necessary ability all magicians should possess.

The first three rituals in this chapter should be performed daily for various reasons. Banishing rituals must be memorized so they can be performed before every magical working without the inconvenience of looking at notes. Practicing these rituals daily makes it easier to commit them to memory. Daily use of these rites can also aid a magician in the development of advanced astral senses and the ability to work with magical energy. Finally, the invoked forces present in all Golden Dawn rituals can act as catalysts for spiritual development and, for this reason, several beneficial changes will occur in one's aura if he or she comes into repeated contact with this Divine current.

Now, before we get to the rituals, a few words should be said about their performance. Some of the words and names of power indicated in the following rituals have to be vibrated (see chapter 2). The actual spelling of these vibrated words will be given in bold capital letters, while the pronunciation of each will follow in parentheses. If you have been practicing vibrating names, you should have no problem using it here. Just make sure you remember that these holy names have built up power over the ages, and their correct vibration is crucial to the success of the ritual.

When performing the rituals, you will have to trace various symbols in the air and visualize them floating before you in their given colors. For now, these symbols can be traced using your pointer finger. I have, however, indicated in the following rituals where you could use a magical tool instead, if you prefer. For example, you can use a dagger to perform the Lesser Banishing Ritual of the Pentagram.

Note: Do not confuse the dagger and wand in the banishing rituals with the Air Dagger and Fire Wand in chapter 4. These last two tools are elemental magical weapons, while the former two are simply extensions of the magician's will.

Any double-edged dagger or wooden rod can be used for this purpose until you make your ceremonial implements. It is still a good idea, however, to at least dedicate them to the service of the Light by saying a few impromptu words over them, such as:

I consecrate and purify thee, creature of steel (or wood), by Divine author-ity, that thou may be a tool in my practice of the magical art and an extension of my will, so mote it be.

It is also a good idea to then pass the tools through running water and the smoke of incense, visualizing any impurities leaving them.

In chapter 2, I recommended you begin a magical diary to record the results of your training exercises. The same holds true for the following rituals. Record any impressions or feelings you have after each session. You will find that on some days the rituals will "feel" a lot more potent or successful than other days. Try to look for patterns that explain when the rituals work well for you and when they don't. The weather, the phases of the moon, and the planetary influences on certain hours can all influence a ritual's success. Also, try not to eat for a couple of hours before a ritual, as the process of digestion inter-feres with magical faculties.

Finally, here is the most important thing to remember before attempting any of these rituals. When you stand in the center of a magical circle, you act with the authority of the Divine, and are in the presence of God. Make sure you act accordingly. Respect for the Divine Energy behind these rituals almost always ensures success, while disregard of its power guarantees failure.

The Lesser Banishing Ritual of the Pentagram (LBRP)

1. Stand in the center of your room, facing east. Imagine that you are a towering figure and that the Earth is a tiny sphere below you. Feel yourself to be the very center of the universe. Then look up into space and imagine a sphere of white brilliance. See this light descend to the top of your head.

2. Reach up with your right hand (or dagger) and pull this white light down to your forehead. When you do this, vibrate the word **ATAH** (ah-tah).

3. Move your hand down your body, feeling the light being drawn down through you in a beam. Touch your breast and move your hand over the groin area, pointing down. Vibrate **MALKUTH** (mahl-kooth) and imagine there is now a shaft of light running through you, connecting the light source above your forehead to the Earth.

4. Touch your right shoulder and imagine that a beam from the shaft of light at the center of your body meets the tip of your finger (or dagger) and moves out from your right side into space. Vibrate **VE-GEBURAH** (v'ge-boo-rah).

5. Touch your left shoulder and imagine that this horizontal beam of light extends to the left and passes out through your left shoulder into space. Vibrate **VE-GEDULAH** (v'ge-doo-lah).

6. Bring both your hands up to your breast and clasp them together, as if praying. Vibrate **LE-OLAHM, AMEN** (lay-oh-lahm, ah-men). You are now standing in the center of a cross of light that reaches to the ends of the universe. (The Hebrew words of power in this part of the ritual, known as the Kabbalistic Cross, translate to: "For Thine is the kingdom and power and glory forever, unto the ages.")

7. Move to the east of your area and trace a large banishing pentagram in the air before you (see figure 3.1). Visualize it glowing in flaming blue light. Bring your hands up to the sides of your head and point both pointer fingers forward (if using a dagger, make sure it is pointing in the same direction as your pointer fingers). Take a deep breath, and thrust both of your arms forward as you advance with your left foot (known as the Sign of the Enterer). At the same time, exhale to vibrate **YOD HEH VAV HEH** (yode-heh-vahv-heh). Feel the energy of this Divine name running through your fingers and into the

FIGURE 3.1

Yod Nun Resh Yod

FIGURE 3.2

pentagram. When this is done, move your left foot back to stand straight again, and put your left index finger to your lips, as if you were telling someone to be silent (known as the Sign of Silence), while you keep your right arm extended.

8. Still touching the center of your pentagram with your finger or dagger and keeping your arm extended, move to the south of your circle. Imagine that the tip of your finger or dagger is creating a white line in the air. When you reach the south, you will have created ninety degrees of a circle of white light, which will connect the pentagram in the east to the one you are about to draw here. Trace the pentagram as before and once again thrust forward into it with Sign of the Enterer, this time vibrating ADONAI (ah-doe-nye). When finished, perform the Sign of Silence, remembering to keep your right arm out before you.

9. Carry the line of white light to the west, and repeat the steps of tracing and charging a pentagram there. This time, vibrate EHEIEH (eh-hey-yay).

10. Carry the line of white light to the north, and trace and charge a pentagram there. This time, vibrate AGLA (ah-gah-lah).

11. Now carry the white line to the east, completing your circle. There should now be four flaming blue pentagrams blazing at the four quarters of your circle of white. Walk back to the center of your circle and turn clockwise to face east.

12. Once again visualize the giant Kabbalistic Cross within you. Extend your arms, forming this cross with your body. Look toward the east and say, *Before me*, RAPHAEL (rah-fay-el). Remember to vibrate the name of this angel. Now

visualize, with your eyes open, that this archangel of Air is standing before you. See him as a towering figure dressed in yellow and violet robes (the colors of Air). Try to feel the breeze of elemental Air in your face.

13. Imagine another presence behind you and say, *Behind me,* **GABRIEL** (gah-bree-el). Again, remember to vibrate the name of this angel. Now visualize, with eyes closed, that this archangel of Water is standing behind you. See him as a towering figure dressed in blue and orange robes (the colors of Water). Try to feel the moisture of elemental Water on your back.

14. Open your eyes and look over your right shoulder while saying, *On my right,* **MICHAEL** (mee-chai-el). After vibrating the name, visualize this archangel of Fire in robes of red and green (the colors of Fire). Try to feel the heat of elemental Fire.

15. Look over your left shoulder and say, *On my left,* **AURIEL** (ohr-ee-el). After vibrating the name, visualize this archangel of Earth in robes of citrine, olive, russet, and black (the colors of Earth). Try to feel the sense of solidity given off by this quarter.

16. Look toward the east again, and contemplate the pentagrams around you, saying, *For around me shines the pentagram* . . . then visualize a brilliant hexagram within your breast, and say, . . . *And within me shines the Six-rayed Star.*

17. To finish this rite, repeat the Kabbalistic Cross in steps 1–6.

This ritual should be memorized as soon as possible and should be practiced every day. Since Divine Names of God, which correspond with each elemental quarter, are used to charge the pentagrams and the archangels of each quarter are called to watch over your area, the circle created in this ritual forms an impenetrable barrier to unwanted magical forces. If this ritual is performed correctly, you can safely proceed with your magical work. Another ritual exists, however, that should be performed along with the Lesser Banishing Ritual of the Pentagram (LBRP) to ensure a complete banishment of unwanted influences. When you feel comfortable doing the LBRP, you can start performing the following ritual immediately after it each day.

The Lesser Banishing Ritual of the Hexagram (BRH)

The Lesser Banishing Ritual of the Hexagram can be performed with either your pointer finger or a wand.

1. Perform the LBRP, making sure to finish the ritual with the Kabbalistic Cross.

2. Extend your arms to form a cross and say, **I N R I.** Then say, *Yod Nun Resh Yod* (yode-noon-raysh-yode). Trace each of these Hebrew letters in the air, from right to left, as they are pronounced (see figure 3.2). Visualize them glowing in the same blue light as the pentagrams in the LBRP.

3. Extend your arms into a cross once more and say, *The Sign of Osiris Slain.*

4. Keep your left arm extended and raise your right arm straight up so it is pointing to the ceiling. Your fingers should be together and extended and your palms should face forward. It should look as if your arms are forming the letter *L.* Look over your left shoulder and say, *L, the Sign of the Mourning of Isis.*

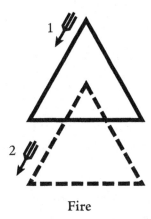

Fire

Earth

Air

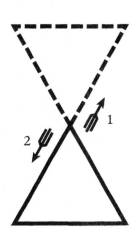

Water

FIGURE **3.3**

5.   Raise both of your arms, palms forward, to form the let-
     ter *V*. The angle between them should be about sixty
     degrees. Tilt your head back and say, *V, the Sign of Typhon
     and Apophis.*

6.   Cross your arms on your chest to form the letter *X*, with
     your palms facing you. Bow your head and say, *X, the
     Sign of Osiris Risen.*

7.   Now form each letter with your arms as it is pronounced,
     L.V.X. Then extend your arms to form the Osiris Slain
     position (cross) and say, *LVX* (lukes).

8.   Keep your arms extended and say, *The Light* . . . after a
     moment, fold your arms on your chest (form an *X* again)
     and continue, . . . *of the Cross.*

9.   Now extend your arms to form a cross again and, as you
     say the following oration, slowly start to raise them up
     to the V position, while gradually tilting your head back:
     *Virgo, Isis, mighty Mother, Scorpio, Apophis, Destroyer, Sol,
     Osiris, Slain, and Risen, Isis, Apophis, Osiris.* Then strongly
     vibrate **IAO** (ee-ah-oh). Your head should be tilted back
     completely by now and your arms should be in the shape
     of a *V*. Become aware of a glowing white brilliance above
     you and say, *Let the Divine Light descend.*

10. Slowly lower your arms to form an *X* on your chest again
     and, as you do so, feel the Divine Light descending
     through your body. Let it energize you for a few moments.

11. Walk to the east of your circle and trace the banishing
     hexagram of Fire in a flaming gold light with your finger
     or wand, over your still shining blue pentagram (figure
     3.3 illustrates the four hexagrams in this ritual). *(Note:
     The hexagrams are not attributed to the placement of the
     elements in the four quarters, but rather to the zodiacal*

placement of each element. Therefore, for this ritual, Fire is in the East, Earth is in the South, Air is in the West, and Water is in the North.) Then charge your hexagram the same way you charged the pentagrams in the LBRP, this time vibrating **ARARITA** (ah-rah-ree-tah).

12. Walk to the south of your circle, tracing a white line in the air like in the LBRP. Trace the banishing hexagram of Earth with your finger or wand and charge this hexagram, vibrating **ARARITA** (ah-rah-ree-tah).

13. Walk to the west of your circle, tracing the line of white. Trace the banishing hexagram of Air, and charge it as before, vibrating **ARARITA** (ah-rah-ree-tah).

14. Walk to the north of your circle, still tracing the line of white in the air. Trace the banishing hexagram of Water, and charge it, once again vibrating **ARARITA** (ah-rah-ree-tah).

15. Complete your circle by carrying the light to the east again. Around you should be your four pentagrams and four hexagrams, with the latter superimposed over the former, blazing in their respective colors of blue and gold at the four quarters of your circle. Contemplate their presence.

16. To close this rite, you can either repeat the Analysis of the Keyword (steps 1–10) or perform the Kabbalistic Cross again. If you are about to perform some type of magic, I recommend you do the Analysis again, as the Divine Light can aid you in your working.

In this ritual, instead of vibrating four different names of power to charge the hexagrams, one word, "Ararita," was used. This is an abbreviation for the Hebrew phrase, "One is His beginning. One is His individuality. His permutation is One."

The importance of the above two rituals cannot be overstressed. To practice evocations, or any other type of ceremonial magic, you have to come into contact with pure and Divine energies. The banishing rituals consecrate your work area and make it a holy place where these energies can descend upon you. These rituals also have to be performed after your magical work is done, to make sure any forces that were invoked or evoked are effectively banished. I also mentioned earlier all of the reasons why you should practice these rituals daily.

The Middle Pillar Ritual

This ritual, while not of a banishing nature, should also be performed daily. The Middle Pillar Ritual activates the energy centers in the human body that correspond to Sephiroth on the Tree of Life through the vibration of Divine Names that correspond to each. In addition to the Divine Name associated with each Sephirah, some magicians like to vibrate the names of angels and angelic orders of each sphere, but this really isn't necessary. I feel vibrating the correct Divine Name of god for each Sephirah three or four times establishes a strong enough link to each energy center.

For the sake of practice, this ritual can be performed right after the LBRP and the BRH, but when you are doing an actual magical ceremony, the Middle Pillar should be used at the point in the working when you need to raise magical energy for some purpose. Later on in the book, you'll see the practical uses for this ritual in performing evocations.

1. Stand facing east and start breathing rhythmically (see chapter 2). Once again, become aware of a glowing sphere of brilliant white light above your head.

2. After contemplating the white sphere above you, vibrate EHEIEH (eh-hey-yay) on your next exhalation and see and feel the sphere glowing brighter. Repeat this two more times or until you feel the white light is blazing like a tiny sun.

3. On your next inhalation, draw a beam of the light down through your head and into your throat center. On the

following exhalation, imagine there is a glowing lavender sphere there. Make sure you still imagine its connection to the white sphere above. As this ritual progresses, you will gradually connect all the centers in your body with this white beam of light. Maintain your rhythmic breathing and on the next exhalation, vibrate **YOD HEH VAV HEH ELOHIM** (yode-heh-vahv-heh el-oh-heem). Repeat this two or more times, each time imagining the lavender sphere growing brighter. Feel it pulsing within your neck.

4. When you feel ready, draw a beam of white light from this sphere to your solar plexus on your next inhalation. Exhale and imagine there is a glowing gold sphere there. Continue your rhythm of breathing, and when you are ready, exhale and vibrate **YOD HEH VAV HEH ELOAH VEDAATH** (yode-heh-vahv-heh el-oh-ah v'dah-aht). Repeat this vibration two or more times and feel the warmth of this sphere within you. Make sure you can visualize all these spheres connected by the white beam of light before you move on.

5. When you are ready, draw a beam of white light from this sphere to your groin center with your next inhalation. Become aware of a violet sphere there and vibrate **SHADDAI EL CHAI** (shah-dai el chai) on your next exhalation. Repeat two or more times, and feel the presence of this sphere within you.

6. When you are ready to continue, draw a beam of white light from this sphere to the area between your feet on your next inhalation. Maintain your pattern of breathing and become aware of a glowing sphere of black energy that is half submerged into the ground. On your next exhalation, vibrate **ADONAI HA-ARETZ** (ah-doe-nye ha-ah-retz). Repeat this vibration two or more times and imagine the sphere is becoming more intense. Feel the energy pulsing there.

7. Continue your rhythmic breathing and contemplate the Pillar standing within you. Make sure you can visualize the five glowing spheres in their respective colors and the white shaft of light connecting them.

8. Now comes the part of the ritual where you can tap the energy of these spheres. On your next inhalation, imagine the energy of the sphere at your feet rise up through your body to the white sphere above you. Then, on your next exhalation, feel this energy explode from this white sphere and fall around your body to the dark sphere below. Repeat seven or eight times. When performing this part of the ritual, you should feel a great amount of energy moving through you. If you wanted to channel this energy into something, a magical tool for example, you could pull the energy up from the center at your feet to your chest area with an inhalation and channel it down through your arms and into the object with your next exhalation.

9. After you have moved the energy through you or used it for some purpose, continue your rhythmic breathing and imagine that the Middle Pillar within you is fading away to a softer glow with each exhalation. You should still feel energized, but a little more relaxed as you are letting excess energy dissipate. The ritual is then completed.

So far we've dealt with banishing and internal power-raising rituals that you can use to consecrate your magical area and to energize yourself. It is true that once your magical area is consecrated, it is ready for ritual use. The forces you are trying to invoke to aid in the ceremony, however, are not present after a banishing ritual. You see, banishing rituals simply clear an area. To perform evocations, or any other ceremonial magic, you have to be able to call on the energies of Divine Providence and the angels of the four quarters. Put simply, you have to invoke necessary energies after you banish unwanted ones. This is the purpose of the following ritual.

The Opening by Watchtower

The Opening by Watchtower was designed by a Golden Dawn Initiate named Israel Regardie as an alternative to the Supreme Invoking Ritual of the Pentagram used by the Order. It contains verses from various mystical sources, including the Oracles of Zoroaster and the Enochian Keys, and is probably the most inspirational ritual opening used in the modern era.

To be truly effective, this ritual should be performed with consecrated magical implements. This includes the four elemental weapons and tablets, the Tablet of Union, and the dagger and wand used in the LBRP and BRH. Chapter 4 contains instructions for constructing these tools and chapter 5 shows you how to consecrate them for use. In the meantime, you should begin practicing this ritual anyway, using much simpler representations of the tools. A feather can be used for Air, a candle or match for Fire, a cup of water for Water, and a wooden bowl of salt for Earth. Temporary elemental tablets can also be constructed. A yellow sheet of construction paper in the east will serve as an Air Tablet, as will a red one in the south for Fire, a blue one in the west for Water, and a black one in the north for Earth. The Tablet of Union (see chapter 4) can be drawn on a white piece of paper.

This ritual is written as if you are using the correct tools or implements. For now, feel free to substitute the above-mentioned implements. Unlike the banishing rituals, this rite does not have to be practiced on a daily basis. Try to become familiar with it, however, as you will be performing it before every evocation. I suggest you write it out on index cards and read from them until you memorize the ritual. Don't worry about trying to learn this ritual, it is really a lot easier to commit to memory than it seems to be at first glance.

1. Prepare an altar in the center of the room. Position the magical tools on the altar so each is closest to the elemental quarter it represents. The Air Dagger should be placed on the east side of the altar, the Fire Wand on the south, the Water Cup on the west, and the Earth Pentacle on the north. In the altar's center, place the Tablet of Union.

Also, make sure the Elemental Tablets are positioned on the correct walls or on stands in the correct quarters. Have your dagger for the LBRP and your wand for the BRH on the southwest and northwest corners of the altar, respectively. Stand behind the altar, facing east.

2.   Pick up the dagger used for the LBRP with your right hand and knock on the altar three times using its handle. Then walk to the northeast of your circle (moving clockwise) and say loudly with meaning, *Hekas, Hekas Este Bebeloi* (hay-kahs, hay-kahs es-stay bee-beh-loy)! This commands all unwanted entities to leave the area because a ritual is about to begin.

3.   Moving clockwise, return to your place behind the altar, facing east. Perform the LBRP and the BRH, making sure to finish with the Analysis of the Keyword. *Note:* When moving around the altar to pick up the magical tools in the following steps, or when pivoting to face the direction of that tool, always move clockwise.

4.   Move clockwise to the south of the altar and pick up the Fire Wand. Pivot and face the Elemental Tablet of Fire in the south. Wave your Wand three times in the air, in front of the tablet, and then raise it. Walk clockwise around the room, keeping the Wand raised, while you say, *And when, after all the phantoms are banished, thou shalt see that Holy and Formless Fire, that Fire which darts and flashes through the hidden depths of the Universe, hear thou the Voice of Fire.*

5.   When you come back to the south, face the tablet and wave the Wand in front of it again three times. Then use the tool to trace a large, blue Invoking Pentagram of Fire in the air before the tablet (see figure 3.4 for the pentagrams used in this ritual). While doing so, vibrate **OIP TEAA PEDOCE** (oh-ee-pay teh-ah-ah peh-doe-kay).

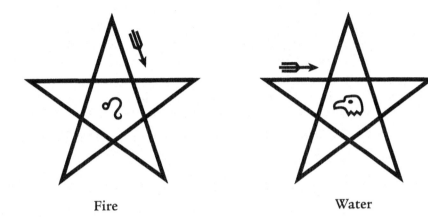

Fire   Water

Air   Earth

FIGURE **3.4**

Then trace the figure of Leo in the center of this penta-gram in red. Point to the center of it with the Wand and vibrate **ELOHIM** (el-oh-heem). Lift the Wand and say, *In the names and letters of the Great Southern Quadrangle, I invoke ye, ye angels of the Watchtower of the South!* Put the Wand back on the altar.

6. Move clockwise to the west of the altar and pick up the Water Cup. Face the Elemental Tablet of Water and wave the Cup before it three times. Lift the Cup above your head, and walk clockwise around the circle saying, *So therefore first, the priest who governeth the works of Fire must sprinkle with the lustral water of the loud resounding sea.*

7. When you return to the west, wave the Cup three times in front of the tablet. Trace a large, blue Invoking Penta-gram of Water using the tool, while vibrating, **MPH ARSEL GAIOL** (ehm-pay-hay ahr-sell gah-ee-ohl). Then trace the blue Eagle head in the center. Point at the center of this pentagram with the Cup and vibrate **ALEPH LAMED AL** (ah-lef lah-med ahl). Hold the Cup above your head and say, *In the names and letters of the Great Western Quadrangle, I invoke ye, ye angels of the Watchtower of the West.* Put the Cup on the altar.

8. Move to the east of your altar, pick up the Air Dagger, and pivot to face the Air Tablet. Wave the Dagger before it three times, and lift it above your head. Walk clockwise around the room with the Dagger in the air while you say, *Such a Fire existeth, extending through the rushing of Air. Or even a Fire formless, whence cometh the image of a voice. Or even a flashing light, abounding, revolving, whirling forth, crying aloud.*

9. When you arrive back at the east, wave the weapon in front of the tablet three more times. Trace a large, blue Invoking Pentagram of Water using the Dagger while vibrating **ORO IBAH AOZPI** (oh-row ee-bah-hah ah-oh-zohd-pee). Then trace a yellow Aquarius sign in the

center. Point at the center of the pentagram with the Dagger and vibrate **YOD HEH VAV HEH** (yode-heh-vahv-heh). Hold the Dagger high and say, *In the names and letters of the Great Eastern Quadrangle, I invoke ye, ye angels of the Watchtower of the East.* Replace the Dagger on the altar.

10. Move clockwise to the north of the altar and pick up the Earth Pentacle. Face the Tablet of Earth and wave the Pentacle before it three times. Then walk clockwise around the circle with the Pentacle high in the air while saying, *Stoop not down into the darkly splendid world, wherein continually lieth a faithless depth and Hades wrapped in gloom, delighting in unintelligible images, precipitous, winding; a black ever-rolling abyss, ever espousing a body unluminous, formless, and void.*

11. When you return to the north, wave the Pentacle in front of the tablet three more times. Then trace a large, blue Invoking Pentagram of Earth using the Pentacle, while vibrating **EMOR DIAL HECTEGA** (ee-mohr dee-ahl hec-tey-gah). In the center, trace a white Taurus sign. Point at the center of the pentagram with the Pentacle and vibrate **ADONAI** (ah-doe-nye). Then hold the Pentacle above your head and say, *In the names and letters of the Great Northern Quadrangle, I invoke ye, ye angels of the Watchtower of the North.* Return the Pentacle to the altar.

12. Move clockwise to the west side of your altar and face east. Over the altar and Tablet of Union, make the following Sign of the Rending of the Veil: Step forward with your left foot and, at the same time, thrust your arms, palms together, forward. Then separate your arms as if you were separating two curtains.

13. Say the following in Enochian, vibrating the words in capital letters: *Ol Sonuf Vaorsagi Goho Iada Balta.* **ELEX-ARPEH COMANANU TABITOM.** *Zodakara Eka Zodakare Od Zodameranu. Odo Kikle Qaa Piap Piamoel Od Vaoan*

(Oh-ell soh-noof vay-oh-air-sah-jee goh-hoh ee-ah-dah bahl-tah. El-ex-ar-pay-hay Co-mah-nah-noo Tah-bee-toh-ehm. Zohd-ah-kah-rah eh-kah zohd-ah-kah-ray oh-dah zohd-ahmehr-ah-noo oh-doh kee-klay kah-ah pee-ah-pay pee-ah-moh-ehl oh-dah vay-oh-ah-noo). This translates to: "I reign over you, says the God of Justice. (Three magical names.) Move, therefore, move and appear. Open the mysteries of creation: balance, righteousness, and truth."

14. Say the following: *I invoke ye, ye angels of the celestial spheres, whose dwelling is in the invisible. Ye are the guardians of the gates of the universe. Be ye also the guardians of this mystic sphere. Keep far removed the evil and the unbalanced. Strengthen and inspire me so that I may preserve unsullied this abode of the mysteries of the eternal gods. Let my sphere be pure and holy so that I may enter in and become a partaker of the secrets of the Light Divine.*

15. Move clockwise to the northeast corner of your circle and face that direction. Say the following: *The visible Sun is the dispenser of light to the Earth. Let me therefore form a vortex in this chamber that the invisible Sun of the spirit may shine therein from above.*

16. Walk clockwise around your circle three times. Each time you pass the east, make the Sign of the Enterer (see LBRP, step 7) in the direction you are moving. This is how you form the vortex mentioned in the last step. After the third time you pass the east and make the Sign of the Enterer, return to the west of your altar and face east. You should feel the vortex of energy in your circle.

17. Make the Sign of the Enterer and say, *Holy art Thou, Lord of the Universe.* Make the Sign of the Enterer again and say, *Holy art Thou, Whom Nature hath not Formed.* Make the Sign of the Enterer one more time and say, *Holy art Thou, the Vast and Mighty One. Lord of the Light and of the Darkness.* Now give the Sign of Silence (see LBRP, step 7).

18. After you perform whatever magical work you want to accomplish and are ready to end the ceremony, say the following: *Unto thee, sole wise, sole eternal, and sole merciful One, be the praise and glory forever, who has permitted me who standeth humbly before Thee to enter into this far into the sanctuary of the mysteries. Not unto me but unto Thy name be the glory. Let the influence of Thy divine ones descend upon my head, and teach me the value of self-sacrifice so that I shrink not in the hour of my trial, but that thus my name may be written on high and my genius stand in the presence of the holy ones.*

19. Now walk around the circle three times, counterclockwise. Give the Sign of the Enterer as you pass the east, in the direction you are walking. The vortex of energy you created earlier should now begin to fade away.

20. Perform the LBRP and the BRH.

21. Say the following, making sure to vibrate the words in capital letters: *I now release any spirits that may have been imprisoned by this ceremony. Depart in peace to your abodes and habitations, and go with the blessings of* YEHESHUAH YEHOVASHAH (yeh-hay-shoe-ah yeh-ho-vah-shah).

22. Knock three times on the altar with the handle of the dagger you used to perform the LBRP and say, *I now declare this temple duly closed.*

Steps 19 to 22 of this ritual are known collectively as the Closing by Watchtower. It should be clear by now, after reading the various orations, the types of links that are created with Divine when this ritual is performed.

Chapter 4 deals with the construction of all of the magical implements you will need to perform evocations. The period of time you spend training your magical senses and learning the rituals in this chapter is ideal for creating your magical tools. This way, you will be ready to perform evocations both spiritually and physically.

chapter four

MAGICAL IMPLEMENTS

The altar represents the foundation of
the Tree of Life: the Sephirah Malkuth.
As such, it is four-sided to represent the
union of the four magical elements in
the physical world.

In this chapter, we'll be dealing with the construction of the implements and tools used in magical evocation, the most basic of which were used before the first grimoires were written. Archaeological evidence shows that magic swords and wands were used by the priesthood in ancient Sumer, who cured people of diseases and performed other magic through the banishment of evil demons and the calling of beneficial spirits. The Sumerians' inspiration for the use of these magical weapons most likely came from their mythology, where gods and goddesses used the same tools to fight the numerous evil beings in their mythos. The centuries-old use of these tools in working with entities indicates just how timeless the practice of evocation is.

The first group of implements that follow are the Golden Dawn Elemental Weapons used in the Opening by Watchtower. I decided to include these specific tools because their symbolism agrees with the nature of the aforementioned ritual. The rituals of evocation, however, found later in the book, are not particularly Golden Dawn oriented and, for this reason, the actual implements used during evocations will contain symbolism that is not peculiar to any specific order. Rather, their designs are drawn from more traditional sources, including grimoires and mythology.

Besides describing the construction and general use of the implements in this chapter, some tools also contain instructions for preliminary purifications that must be performed during or before construction. This should not be confused with the process of charging and consecrating magical tools for use, which is given in chapter 5. The preliminary purifications given in this chapter are of the type that cannot be performed after a tool's completion, and these tools still have to be charged and consecrated for use, once constructed. It is helpful to think of purifications as banishing rituals that rid the tool of unwanted influences, and of consecrations as invoking rituals that attract favorable influences to a tool.

I've tried to make the instructions for constructing these tools as easy to follow as possible. In many cases, the construction of a tool requires nothing more than painting it. This is true in the case of the Air Dagger, for example, where the handle has to be painted. Some of the magical implements described in this chapter can also be bought, ready-to-use. An example of this type of magical tool is the censer or incense burner.

No special skills are necessary for the construction of the following implements. You may need to use power tools, however, if certain objects cannot be found in naturally correct shapes. For example, you might not be able to find a flat disc of wood to make an Earth Pentacle. If you are not familiar with the use of power tools, ask an experienced friend to help you or order a precut piece of wood in the correct shape and dimensions at a local hardware store. This last option is one you will have to use when constructing your magic mirror. I don't recommend trying to cut your own disc of glass, unless of course you happen to have that skill.

When painting the Elemental Weapons and Tablets, you will have to use what are known as "flashing colors." These are complementary colors that appear to "flash" when near each other. For example, if you paint something red, and then paint a green letter over it, the contrast of the two complementary colors will produce a flashing effect. This technique doesn't always work very well, because painting a color over another one usually causes the second color to appear dull. A slightly complicated, yet effective procedure can be used to prevent this from happening: paint your symbol or letter in white first, then paint over it with your flashing color. In the above example, you would paint your object red, let it dry, then paint your chosen letter in white, let it dry, and finally paint over the white letter in green. This technique is complicated because the flashing letters and symbols are often small, and painting over them is difficult. Use whichever method works best for you.

The basic colors (red, green, blue, orange, etc.) of any brand of acrylic paint should flash when contrasted with their complementary colors. However, colors with other than basic names (like candy apple red or magenta) usually don't flash. Before painting your tools, you should test this effect with whatever paint you buy. Some brands of paint tend to create softer looking colors, which do not flash as distinctly as other brands do. Try to get bright colors of acrylic paint to achieve the best effect. When painting your tools, remember to use a coat of white primer first, to make sure your colors "take." Also, coat your finished tools with a lacquer-type finish, to make sure their color doesn't start to dull with use.

Golden Dawn Elemental Weapons

Following are the four Golden Dawn Elemental Weapons: Fire Wand, Water Cup, Air Dagger, and Earth Pentacle, given in order of appearance in the Opening by Watchtower.

FIGURE 4.1

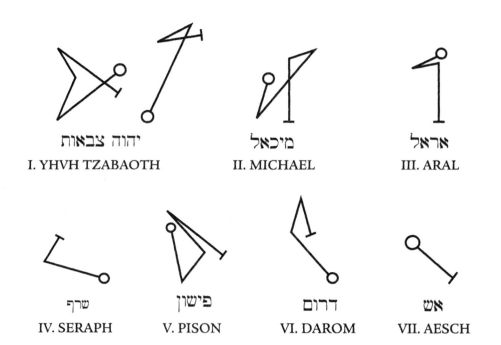

 יהוה צבאות

I. YHVH TZABAOTH

מיכאל

II. MICHAEL

אראל

III. ARAL

שרף

IV. SERAPH

פישון

V. PISON

דרום

VI. DAROM

אש

VII. AESCH

FIGURE 4.2

The Fire Wand

This tool is used to invoke the power of Fire in a ritual. For the purposes of evocation, it will only be used, along with the other three Elemental Weapons, in the Opening by Watchtower Ritual. The Golden Dawn's instructions for the Fire Wand stated that it had to be a phallic shape, and that it should contain a magnetized wire in its center, extending from both ends or tips of the implement. This last detail would make the Fire Wand the most difficult Elemental Weapon to construct, if the presence of this wire was really necessary. However, the only wand that should contain a magnetized wire is the Magic Wand, described later in this chapter, because unlike the Fire Wand, this wand performs an energy-channeling function (still, the presence of the magnetic wire in the Magic Wand is only recommended, not required). Therefore, the construction of the Fire Wand is actually quite simple.

To make a Fire Wand, you will need a wooden dowel that is about eight to ten inches long, and anywhere from ½ to ¾ inches in diameter. For the tip of the wand, you will have to find a cone or acorn-shaped piece of wood (see figure 4.1). Try your local hardware store, as they usually have decorative furniture molding in this shape. If you can't find one, use a lathe to make one or carve an approximation of it from some soft wood. Use a file and coarse sandpaper to make the cone smooth if you are carving it yourself.

To connect the cone to the dowel, do not use screws, staples, or nails. Get a thin wooden peg and measure its diameter. Then use a drill bit of the same diameter to drill one ½-inch-deep hole in the center of the end of the dowel and one ½-inch-deep hole in the center of the flat end of the cone. Cut the thin peg to a one-inch length and insert it into the dowel. Then fit the cone onto the exposed end of the peg. If the fit is not tight and there is a space between the two pieces of your wand, try cutting a thin slice off the end of the connecting peg. Once you get a tight fit, apply wood glue to the bases of the dowel and cone and to the peg and reconnect the cone to the dowel. Let the glue dry.

When the wand is dry, paint it with a couple of coats of white primer, letting each coat dry. You should apply as many coats as is necessary to make sure the wood is no longer absorbing the paint. When your final coat is dry, paint three elongated Hebrew Yuds on the cone in bright yellow. Also, paint four ¾-inch-wide yellow bands around the shaft of the wand. The first one should be painted where the cone meets the shaft, the last at the other end of the shaft, and the other two should be equally spaced, as if the wand were sectioned into thirds (again, see figure 4.1).

When the yellow paint dries, paint the rest of the shaft, and the three spaces between the yellow Yuds, a bright red. On these areas of red, you can then paint the seven Hebrew words and sigils in bright green, using one of the methods described earlier in this chapter (these symbols appear in figure 4.2). In addition, you can also add a magical motto or magical name (if you have one) to the wand to personalize it. Paint this in green on one of the red areas of the wand as well.

When the paint on your wand is dry, coat the entire weapon with a clear lacquer finish and let it dry. Your Fire Wand is ready to be consecrated (see chapter 5). Wrap it in a piece of silk that is either red (symbolizing Fire), black (in this case, symbolizing the absorption of energy), or white (symbolizing the Light), and put it away until you are ready to consecrate it.

Silk handkerchiefs of various colors and sizes can be purchased in stage magician supply stores. Silk is a natural magical insulator and should be wrapped around all stored magical items. In the case of the four Elemental Weapons and Tablets, you should try to store each in its appropriate elemental color. When silk of the correct color is not available, you can use either black or white for the above-mentioned reasons.

The Water Cup

This magical implement is used to invoke the forces of elemental Water. The construction of this elemental weapon is really nothing more than painting an already existing cup with a stem. The design of the cup is supposed to resemble the appearance of a crocus flower, with the top of the cup flaring out, although a cup with a rounded top will work equally well. The Hebrew names and symbols appropriate to the Water Cup are drawn in what appear to be eight crocus leaves. I recommend you try to find a new wooden goblet of this general shape in a variety store, or a used one in either a garage sale or thrift store. Metal

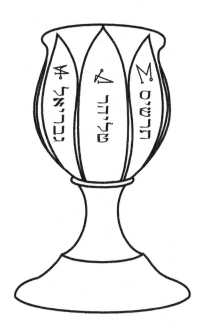

FIGURE 4.3

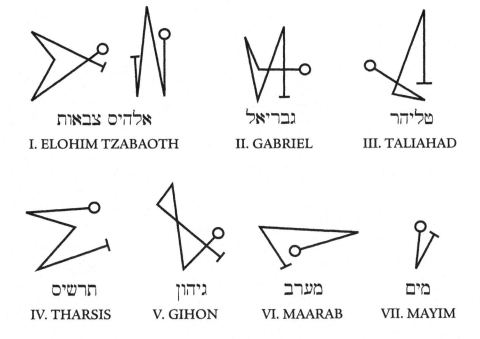

אלהים צבאות
I. ELOHIM TZABAOTH

גבריאל
II. GABRIEL

טליהר
III. TALIAHAD

תרשיס
IV. THARSIS

גיהון
V. GIHON

מערב
VI. MAARAB

מים
VII. MAYIM

FIGURE 4.4

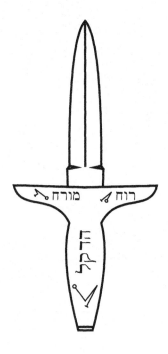

FIGURE 4.5

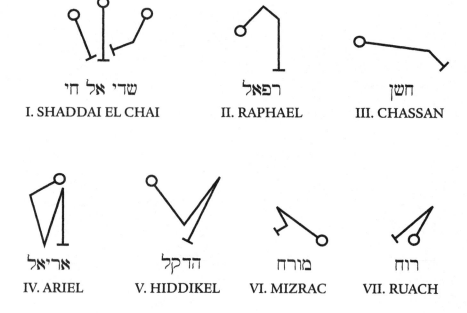

שדי אל חי

I. SHADDAI EL CHAI

רפאל

II. RAPHAEL

חשן

III. CHASSAN

אריאל

IV. ARIEL

הדקל

V. HIDDIKEL

מורח

VI. MIZRAC

רוח

VII. RUACH

FIGURE 4.6

goblets are sometimes difficult to paint, and glass ones break easily, so wood is really the best choice. If you obtain a used cup, you should pass it under running water and visualize any impurities in it being washed away.

If your cup is made of wood, it may contain a layer of clear finish. Remove this with paint remover before proceeding. Then, in light pencil trace eight equal petals around the cup, with slightly rounded triangular tops (see figure 4.3). Draw the bottoms of these leaves so they terminate where the rounded part of the cup ends and the stem begins. Paint over this pencil outline of the leaves in white primer, and let it dry. Then paint the outlines of the leaves in bright orange, making sure you distinctly form eight separate leaves as shown in figure 4.3.

When the orange outline dries, carefully paint the insides of the leaves with white primer. When this dries, paint the white leaves a bright blue color. Then paint one of the seven sigils in figure 4.4 at the top of each leaf in orange (again see the Water Cup in figure 4.3), leaving the eighth leaf blank. Also paint the Hebrew words that correspond to each sigil in orange along each leaf. On the eighth leaf, you can put your magical name or motto in orange to personalize your Water Cup. Leave the stem of the cup and the spaces between the tops of the leaves unpainted.

When your Water Cup is dry, coat it with clear lacquer and let it dry. Then wrap it in a piece of blue, black, or white silk and store it until you are ready to consecrate it.

The Air Dagger

This weapon is used to invoke the elemental forces of Air. Like the Water Cup, there is really no construction involved in the creation of an Air Dagger, only painting. The Air Dagger has to be double-edged, and should have a T-shaped handle (see figure 4.5), which is large enough to fit all the words and sigils given in figure 4.6. The ideal dagger will have a handle made of wood, as this is easiest to paint on. Metal handles are often bumpy, and some wooden handles contain carvings, so no matter what type of material your dagger's handle is made of, make sure its surface is smooth. The place where the hilt of the dagger meets the cross bar to form a T is the only place where simple designs or lines should be found. If there are any lines or designs on the other parts of the handle, then you should look for another dagger.

Once you find a dagger that is suitable for use, wrap the entire blade with tape to protect it from paint, and give the handle a necessary number of coats of white primer. When these have dried, paint the handle a bright yellow, and let it dry. Using whatever method you prefer, paint the seven Hebrew words and sigils on the hilt in bright purple. If you have a magical motto or name, paint this on the handle as well. When the handle is dry, apply a coat of clear lacquer finish. Do not remove the tape from the blade until the lacquer is dry.

Your Air Dagger is then ready to be consecrated. Wrap it in a piece of yellow, black, or white silk and put it away until you are ready to do so.

The Earth Pentacle

This last Elemental Weapon is used to invoke the forces of Earth. The construction of the Earth Pentacle can be very difficult if you can't find a precut disc of wood or have someone cut one for you. Visit you local hardware store and ask if they have any flat discs of wood about five inches in diameter and ½ to ¾ inch thick. You might be able to obtain a pre-cut one that was made for some decorative purpose, or have the store order or cut one for you. If the above methods fail, try to get someone who is skilled with power tools to help you cut a disc, unless you are skilled yourself.

To cut your own disc, take a square piece of wood (about 5 x 5 inches) and use a compass to draw a circle that is about five inches in diameter (it should just fit within the square). Clamp this square of wood onto a workbench so half of it clears the edge of the bench. Carefully cut out this half of the disc using a power tool known as a jigsaw. Then turn the wood around and clamp it again, so the uncut half of the circle clears the edge of the bench. Carefully cut out this half of the disc using the jigsaw. When you are finished, sand the edges of the disc so it is smooth all around.

Paint the entire disc with as many coats of white primer as are necessary. When the paint dries, use a compass to lightly trace a second circle on the disc that starts about ½ inch from its edge (see figure 4.7). Divide this inner circle into four equal pieces, using two lines. Then draw a hexagram whose top and bottom points match the vertical line you drew. Draw two more lines, this time diagonally. Erase the vertical and horizontal lines, so you are left with a hexagram that is drawn over an X, which divides the circle into four equal pieces.

FIGURE 4.7

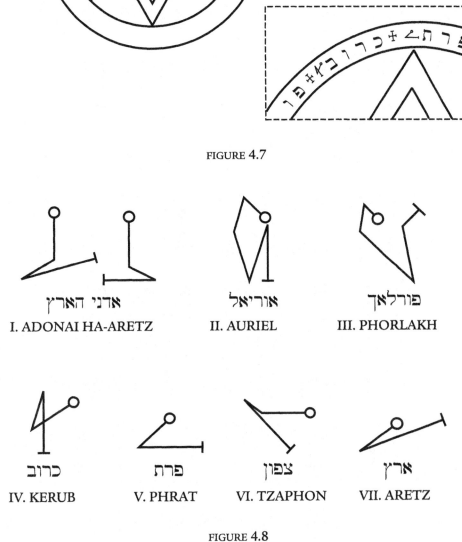

I. ADONAI HA-ARETZ

II. AURIEL

III. PHORLAKH

IV. KERUB

V. PHRAT

VI. TZAPHON

VII. ARETZ

FIGURE 4.8

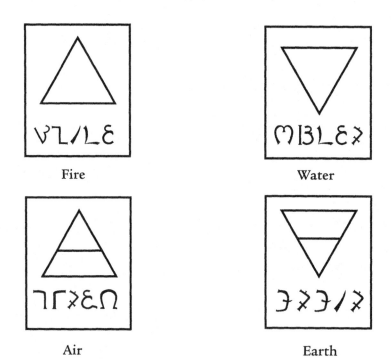

Fire

Water

Air

Earth

FIGURE 4.9

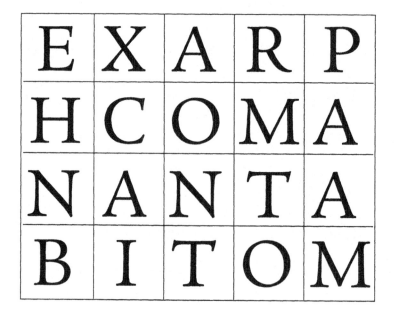

E	X	A	R	P
H	C	O	M	A
N	A	N	T	A
B	I	T	O	M

FIGURE 4.10

Draw the same-sized circle on the other side of the disc, and repeat all the above steps so you have an X and a hexagram on both sides of the disc. Make sure the hexagram and X on each side are carefully aligned with the ones on the other side. In other words, if the disc were clear, and you were to look through it, the lines on both sides would perfectly overlap, forming one X and one hexagram.

Mark the points of the hexagrams with a pencil on the white borders of each side of the disc so you can recreate them later. In the following steps, you will have to paint the four quarters of the X in the inner circle. Do not paint over the white border of the outer circle. Paint the top quarter of the X on each side of the disc citrine, which is a combination of orange and green. Then paint the right quarter of each side of the disc olive, which is a mixture of green and violet. The bottom quarter of each side of the disc should be painted black. Finally, paint the left quarter of each side of the disc russet, which is a mixture of orange and violet.

Using a pencil, connect the points you marked on the white border of the disc to trace a hexagram again. About ⅜ inch within this hexagram, trace another, parallel hexagram. When you paint the space between these lines white, the result will be a ⅜-inch-thick white hexagram. Repeat this process on the other side.

Erase the markings on the white borders of your disc, or better yet, paint another white coat over them. When this has dried, paint the Hebrew names or words given in figure 4.8 in black around the white border of each side of the disc. After each name or word, paint the appropriate sigil followed by a cross. The crosses will separate each pair of words and sigils (see the enlarged portion of figure 4.7). Also, paint your magical name or motto in black, followed by a cross. Try to space the words and sigils so they completely circle the white border. When the paint is dry on your disc, apply a clear coat of lacquer finish.

Your Earth Pentacle is finished. Wrap it in black or white silk and put it away until you are ready to consecrate it.

Elemental Tablets

The Elemental Tablets of the four quarters given below each consist of a symbol and an Enochian word (in the Enochian Alphabet) for the element. These words are also found on the Tablet of Union (in English characters), which remains on the altar. Some magicians like to place complete Enochian Watchtower Tablets in the four quarters of the circle, however these are difficult and time-consuming to make. The simple tablets given here more than adequately represent each element, and can be constructed rather quickly. As I mentioned earlier, they should also be painted in flashing colors.

The four Elemental Tablets and the Tablet of Union can be painted on either heavy poster board or thin plywood. For the Elemental Tablets, use whatever size board or plywood you feel is most convenient to hang on your wall, although I don't recommend anything smaller than a square foot. The Tablet of Union should be whatever size best fits on your altar.

The Elemental Tablet of Fire

This Elemental Tablet represents the invoked forces of Fire. It should be hung in the south wall of your personal temple and is the focus of your invocation of the power of this quadrangle in the Opening by Watchtower. The board used for this tablet should be painted a bright green, while the symbol of Fire and the Enochian word for Fire, "Bitom," should both be painted upon this board in bright red (see figure 4.9). It is a good idea to trace the Enochian letters before painting them, as their shapes are powerful magical symbols and should be drawn correctly.

When your Tablet of Fire is completed, apply a coat of protective finish to it (if it is made of wood) and hang it on the appropriate wall of your personal temple. To consecrate it, simply perform the Opening by Watchtower after your Elemental Weapons are constructed and consecrated.

The Elemental Tablet of Water

This tablet represents the invoked forces of Elemental Water. The board used for this tablet should be painted bright orange, while the symbol of Water and the Enochian word for Water, "Hcoma," should both be painted bright blue (see figure 4.9). Remember to apply a coat of protective finish if your tablet is made of wood and hang it up on the west wall of your temple when it is completed. Like all Elemental

Tablets, the Tablet of Water is consecrated by its use in the Opening by Watchtower.

The Elemental Tablet of Air

The Tablet of Air represents the invoked forces of the Eastern Quadrangle. Paint the board used for its construction a bright purple. The triangle, which is the symbol of Air, should be painted in bright yellow, and the line going through this triangle should be painted purple. Also, paint the Enochian word for Air, "Exarp," in bright yellow (see figure 4.9). Hang this tablet on the east wall of your temple.

The Elemental Tablet of Earth

This tablet represents the invoked forces of Earth. The board used for this tablet should be painted silver. Paint the triangle that represents Earth in a flat black, and paint the line that goes through it silver. The Enochian word for Earth, "Nanta," should be painted flat black (see figure 4.9). Hang your finished Tablet of Earth on the north wall of your temple.

The Tablet of Union

The Enochian names of the elements used on the above tablets are given in the Enochian Alphabet for a specific reason. As mentioned earlier, the Enochian characters have a mystical power of their own and writing the names of the elements in this alphabet helps invoke their presence in the tablets. They act as talismans, drawing elemental energy to your temple.

The Enochian names of the elements given in the Tablet of Union, however, are not given in the Enochian Alphabet. This is because the Tablet of Union is supposed to symbolize the union of the four elements on the altar. By putting the names of the elements in an alphabet you can recognize, you help to establish their presence through recognition. When you see "BITOM" (the names are drawn in capitals on the Tablet of Union), you should remember, at least subconsciously, that Fire is one of the elements present on your altar. The same is true for the other names.

The Tablet of Union should be made out of the same material used in the construction of your Elemental Tablets. Cut the board to a rectangular shape so the ratio of its length to its width is five to four. For example, use a board that is ten inches long by eight inches wide. The board should be white, so if you are using wood instead of white poster board, be sure to paint it so.

Divide the board into twenty equal squares. The resulting grid should be five squares long across the length of your board, and four squares deep across its width. In the above example of a board that is 10 x 8 inches., the squares would each be two inches square. The lines making up the squares should be painted black.

Paint the word "EXARP" in capital letters across the first row of squares, in yellow letters. On the second row, paint "HCOMA" in blue. "NANTA" should be in black on the third row, and "BITOM" should be painted in red on the fourth row (see figure 4.10).

If you are using wood, apply a coat of lacquer finish and let it dry. Your Tablet of Union is then ready for use. Like the Elemental Tablets, it is charged during the Opening by Watchtower. Lay it flat on the center of your altar so you can read it when standing in the west, facing east.

Other Implements

The Magic Wand

The Magic Wand represents the absolute will of the magician. This potent magical tool can be used for almost any energy-directing purpose, which includes healing, the charging of other magical implements, and the evocation of benevolent spirits (malevolent entities should be called with the magic sword, which is explained below).

Over the years, several occult orders and magical traditions have each developed specific magic wands for use in their rituals. The Golden Dawn, for example, had a different wand for every officer of the temple (Chief Adept's Wand, Praemonstrator's Wand, etc.), a wand for invoking the element of Fire (described above), and a Lotus Wand that was useful for many types of rituals. As a solitary ceremonial magician, however, you only need to concern yourself with the already explained Fire weapon and one other type of wand, the Magic Wand, which is used for directing energy and willpower.

This Magic Wand does not have any particular elemental or astrological correspondences associated with it, therefore it is free of markings, unlike the Golden Dawn Elemental Weapons. The Magic Wand is a universal instrument, and can be used within the context of any magical current or tradition. When charging and consecrating your other magical weapons, this is the instrument you will use to help you direct energy and Divine Light into them.

The type of material selected for a Magic Wand's construction is very important. Over the centuries, several different woods have been used and the ones recommended here are used almost universally. Ash, cane, elder, hazel, oak, and willow are all ideal choices for this magical implement's construction. A rod of elder has a pith in the middle that can be removed, and cane has an already hollow center, making these two woods ideal for creating a Magic Wand with a magnetized wire running through it. The other types of wood mentioned can be used for creating wands because of their individual occult properties, which were known by magicians hundreds of years ago who included them in their grimoires.

To create your Magic Wand, cut a shaft from one of the appropriate trees mentioned above. Make sure the shaft is straight and does not have any offshoots or branches growing off it. According to tradition, the length of the Magic Wand should be equal to the distance between the tip of your extended pointer finger and your shoulder. I recommend you try to follow this tradition because it seems to have a personalizing effect on the Magic Wand. Remember, you can use your pointer finger and entire arm to project energy (as you did in the BRH when you traced hexagrams). Therefore, measuring your wand to the length of your arm and extended finger creates an extension of yourself, and consequently, of your will.

It is a good idea to cut off more than the desired length of your wand so you can cut it straight and to the right length at a bench. Make sure the tips of the wand are both flat and level. Once your Magic Wand is cut to the right length, the next step is to remove the bark and sand your implement until it is smooth. If you are using cane or elder wood, then you could easily insert a magnetic wire into the center of the shaft. The presence of the magnetic wire is not really necessary in a Magic Wand, however, as its ability to channel your magical energy depends only upon your consecration of the instrument.

If you decide to use a magnetic wire, you will have to first create one. Take a piece of steel wire and repeatedly run a magnet along it in one direction. After a few minutes, test the wire by trying to pick up a paper clip with it. If it is ready, the next step is to carefully insert the wire into your Magic Wand. Make sure the wire is perfectly straight to make this easier. After you insert it into the center of your shaft of wood, you can then paint the wand with a coat of clear finish to protect the wood.

When your Magic Wand is complete, wrap it in either a white or black piece of silk and either put it away until you are ready to consecrate it or perform a minor consecration like the one given in chapter 3, and use it when you perform the BRH. Using your Magic Wand in this ritual will help you develop a link to the tool, which will make its consecration much more effective.

The Magic Sword and Dagger

Like the Magic Wand, the Magic Sword is an instrument of the magician's will. This weapon, however, can also be used to enforce one's will when dealing with evil or malevolent entities, making the Magic Sword an instrument of authority. The Magic Sword is of the same nature as the dagger used in the LBRP. This dagger helps the magician banish unwanted influences with a Divine Authority, and in some ways the Magic Sword can be seen as an enhanced version of this dagger.

Since the dagger and Magic Sword are similar, both weapons will be dealt with here. There is no physical preparation necessary for either of these tools. A magician only has to find a suitable sword and dagger for use. The Magic Sword should be light enough to hold comfortably for an extended period of time. The length of the blade and the type of handle on the sword are not important, but the blade should be double-edged. The same holds true for the dagger. Like the Magic Wand, these two weapons do not belong to any particular tradition, and because of their universal nature should be free of any markings. The power of these two weapons to help a magician enforce

FIGURE 4.11

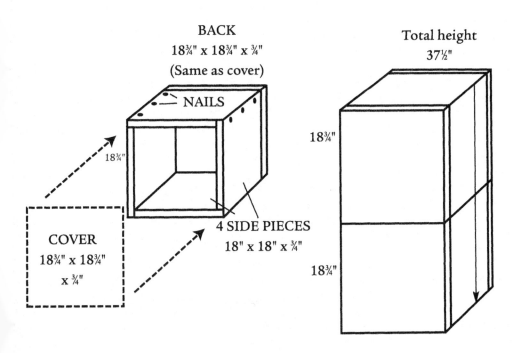

BACK
18¾" x 18¾" x ¾"
(Same as cover)

NAILS

18¾"

COVER
18¾" x 18¾"
x ¾"

4 SIDE PIECES
18" x 18" x ¾"

Total height
37½"

18¾"

18¾"

FIGURE 4.12

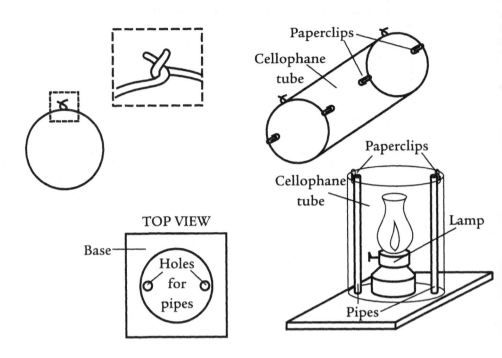

TOP VIEW

FIGURE 4.13

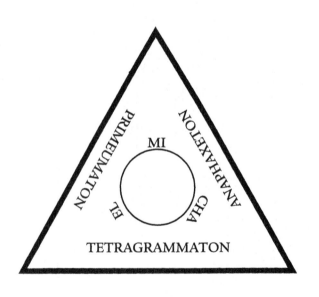

FIGURE 4.14

his or her will is bestowed upon the sword and dagger when they are consecrated.

The dagger you choose can be consecrated in a minor fashion, as given in chapter 3, and can be used for performing the LBRP. When not in use, it should be wrapped in white or black silk. The sword should also be wrapped in white or black silk and put away to insulate it against any negative influences. Both tools should be consecrated with the ritual given in chapter 5 before they are used for evocations.

The Robe

This robe is a very important implement for the magician to have, as putting on a robe helps one get into a magical state of mind. Your everyday clothes remind you of everyday things, and this mundane mood is not desired in a magic temple. Putting on the robe signifies that you are leaving the physical world and entering a magical one, where all planes and dimensions coexist. For this reason, the robe should only be worn when performing magic.

To make the robe, you will need a large piece of cloth that is as wide as your height minus one foot, and twice as long as your height minus about 1½ feet. For example, if you are six feet tall, your selected piece of cloth must be five feet wide and about ten and one-half feet long. Use a material you think you will be comfortable in which is colored either white or black. You should be able to find a wide enough roll of cloth at a local fabric store. If you have difficulty finding the color or size of cloth you need, however, try asking a company or factory that produces large amounts of fabric for clothes manufacturers to sell you some.

Fold this cloth, inside out, so it becomes half its length. Using the above example, it should now measure 5' x 5'3". Enlarge the pattern given in figure 4.11 and trace it on the cloth using a cloth marker. The opening for your head could contain a slit on one side, as shown, making it easier to put on. Make sure the arms of the robe begin where your arms begin. The best way to make sure of this is to lay down on the fabric and have a friend mark where your shoulders and armpits are. Also, make sure the opening you are tracing for your head is at the center of the robe.

Keeping the cloth folded and both sides firmly together, cut out the pattern in figure 4.11. Remember, only one side of the robe should contain a slit below the head opening. After your robe is cut out, sew along the sides and the bottoms of the arms. When this is done, turn your robe inside out, so the stitches are no longer visible. Your robe is then ready to be worn during magical rituals.

The Ring

Just as the robe helps a magician enter a magical state of mind, the ring also alerts one's subconscious that mystical work is about to be done. It is worn on the projecting finger (the pointer finger of your right hand if right handed, or left if you are left handed). Its design can be a simple band; a band with a stone that has some meaning to you, like a birthstone; or a band with a gem with an occult significance you feel relates to you (see *Cunningham's Encyclopedia of Crystal, Gem & Metal Magic,* originally published by Llewellyn in 1988 and revised in 2002, for the occult properties of the components of your ring). You could also use a ring with a special esoterical design. Many occult supply stores carry a Ring of Solomon, which is illustrated in the *Goetia.* Wearing this ring would be especially appropriate when performing evocations.

Whatever ring you choose, start wearing it with your robe when practicing your rituals. When you put your robe and ring on, make sure you contemplate the significance of the change of consciousness that is taking place. You should feel prepared to do magic and any troubles or everyday thoughts should be left behind.

The Altar

The altar represents the foundation of the Tree of Life: the Sephirah Malkuth. As such, it is four-sided to represent the union of the four magical elements in the physical world. Also, the altar should be a double-cube, with the top cube representing the macrocosm and the bottom cube representing the microcosm, as in the hermetic axiom, "As above, so below." This makes the altar a symbol of Divine Work being done on the physical plane. It is placed in the center of the magic circle, where the vortex of energy raised in the Opening by Watchtower is also centered.

Your altar should be tall enough for you to pick up an item off of it without bending over. About three feet tall is good for most people. A small cubical cabinet could be used for an altar, but building one yourself is recommended if you can get access to a table or circular saw.

To make a true, double-cube altar, you have to cut twelve squares from ¾-inch-thick plywood using a table or circular saw. Eight should be 18 x 18 inches, and four should be 18¾ x 18¾ inches. Glue and nail four of the former and two of the latter together to form a sealed cube as shown in figure 4.12. Then glue and nail the four remaining 18 x 18 inch wood squares and one of the 18¾ x 18¾ inch squares to form an open cube (again, see figure 4.12). When the glue is dry, nail the open cube to the top of the closed one, making sure the two are aligned. Finally, nail the last square to the top cube to seal the opening. You should now have a true, double-cube altar.

Apply a coat of white primer to your altar and let it dry. Then paint the entire altar a flat black. When the paint dries, make sure to apply a coat of lacquer finish to the entire altar. Do not add coasters, legs, or wheels to your altar, as it is supposed to form a foundation with the earth. Begin using the altar in your rituals as soon as it is finished.

The Magic Circle

The magic circle is created in the banishing rituals and further strengthened in the Opening by Watchtower. If you perform these rituals correctly, you should become aware of the glowing circle of white around you. Therefore, it is not really necessary to have a physical representation of the magic circle. If it would make you comfortable to have an idea where this boundary is at all times, however, you can have a physical representation of the magic circle on the ground around you.

One way to draw a magic circle is to use chalk. This is only good if you are performing evocations in a basement with a concrete floor, however. If you are in a room in your house, a good way to represent the magic circle is to use a white ribbon or rope. Outdoors, you have two excellent options available to you. Rocks can be arranged in a circle around you, or you can dig a circle into the ground using a trowel. Remember, these are just representations of the location of your real magic circle of light. Do not worry about this physical circle being open in areas (rocks do not make a tight seal), or about it not being the same as one you found in a grimoire (all the names and symbols drawn there are not necessary). The only circle that really matters is the psychic one you create through rituals.

The Lamp

The lamp used in evocations is a common oil lamp. It can be placed in a few different areas of the temple, depending upon what type of evocation you are performing. A lamp is only necessary when performing evocations to the physical plane, and can be substituted with candles when you are evoking to the astral.

To perform an evocation to the physical plane, you will have to fill your temple with light that corresponds to the sphere your chosen entity comes from. For example, if you are evoking an angel from Tiphareth, you will have to fill your room with a golden-yellow light. The lamp has to therefore be covered with some type of colored filter, which will cause it to give off the correct color light.

Instead of making a permanent filter for each color you might end up using, I have designed a simple filter that allows you to exchange different colors of cellophane plastic. Sheets of this can be bought at a craft store in virtually every color. To make the filter, take thick steel wire and bend it into a circle that is four times as wide as the diameter of the base of your lamp (see figure 4.13). Twist the ends of the wire into little hooks, and connect them together, so you have an interlocked wire circle. Depending on the diameter of your lamp, this circle could be up to two ft. wide.

Cut a square of plywood that is three or four inches. wider than your wire circle. Then, get two thin pieces of copper pipe that are six inches longer than the height of your lamp. Measure the diameter of these pipes, and get a drill bit of the same diameter. Lay the wire circle on the wood and trace its shape in pencil. On opposite sides of this pencil circle, draw two dots indicating where you will be drilling (see figure 4.13). Drill two holes at these points. The holes should just fall within the circle, and should just touch it (again, see figure 4.13). Stand the two pipes in their respective holes.

To use the filter, wrap a piece of cellophane into a tube that is as tall as the two pipes and as wide as the wire circle. Allow at least one inch of cellophane to overlap and cut any excess off. Attach the wire to one end of the tube using two paper clips, making sure one clips the overlapping part of the tube. Put two paper clips on the other end of the tube so they are aligned with the top ones, thereby holding the tube together (again, see figure 4.13). Slide the tube over the pipe stand so the wire circle is on top and slide the two top paper clips into the pipes to secure the tube to the pipe stand. In other words, the top paper clips should each be attached to a pipe, the wire circle, and the edge of the cellophane tube. You can then make different colored tubes and clip them to the stand whenever you need them.

To use your lamp with a filter, do not set it too bright. Because of its large diameter, the tube of cellophane described above should not be affected by the heat of your lamp. Turning your lamp to its brightest setting, however, may cause the cellophane to melt if you leave it on too long. Set your lamp at about half-wick, and always use the glass cover that goes with it.

The Censer

The censer is nothing more than an incense burner. This can be store-bought, or a bowl filled with sand. To use it, light a piece of self-lighting charcoal (ask for it where incense is sold) and lay it in the censer. On top of this burning charcoal, you can throw whatever incense you will be using.

In evocations to the astral plane, the censer should go on the altar. The corresponding incense will then help the magician attune to the sphere he or she will be skrying. In evocations to the physical plane, two censers will be necessary. One will go on the altar and one will go in the Triangle of the Art (see below). The same incense, corresponding to the entity, should go in both.

The Triangle of the Art

The Triangle of the Art is also known as the Triangle of Manifestation because this is where evoked entities appear. In evocations to the astral plane, this triangle is placed flat upon a table outside of the magic circle, and either a crystal or magic mirror is placed on top of it, while in evocations to the physical plane, it is placed on the floor outside the circle.

The right-side-up equilateral triangle is the symbol of Fire, which is the element of manifestation. The name of the Archangel of Fire, Michael, along with three Divine Names are written inside the Triangle of the Art. The three sides of the triangle also symbolize the three planes a manifestation has to go through: mental, astral, and physical. The circle in the center of the triangle helps to contain the manifestation of an evoked entity. Both the triangle and circle are physical representations of a psychic triangle and circle, which the magician traces over them with his or her Magic Wand during an evocation.

To create the Triangle of the Art, cut an equilateral triangle (each angle is sixty degrees) out of plywood. Each side should be three feet long. Apply a few coats of white primer, and when it dries, paint the entire triangle iridescent white. The edges of the triangle should be painted black. The three Divine Names **PRIMEUMATON, TETRA-GRAMMATON**, and **ANAPHAXETON** should be painted on their respective sides in black capital letters (see figure 4.14). The name Michael is divided into three syllables: **MI, CHA**, and **EL**, which should be painted on their respective corners in red. Finally, the circle in the center should be painted green, which is the flashing color of Fire. Apply a coat of clear lacquer finish when the paint on the triangle is dry.

The Crystal

Evocations to the astral plane can be performed with either a crystal or a magic mirror. If you have been skrying with a crystal, then you should also use the crystal for astral evocations. Before using your crystal, purify it by holding it under running water and visualizing any impurities leaving it. Then dry it and consecrate it using the ritual in chapter 5. Find a table that is about the same height as your altar, position it outside of your magic circle (see chapter 5), and lay the triangle on top of it. Your crystal should be placed on its stand in the center of the circle within the Triangle of the Art. As always, any lighting in the room should be behind you, to make sure no reflections are seen in the crystal.

The Magic Mirror

This is the preferred tool for evoking entities to the astral plane. Since it is only one-sided, the magic mirror is not as prone to casting reflections of your temple as is the crystal. The magic mirror is an ideal doorway to the astral plane and as mentioned in chapter 3, it costs much less to build a magic mirror than it does to buy a crystal. In this section are instructions for making a magic mirror and stand.

To construct a magic mirror, you will have to acquire a pre-cut disc of glass. Most glass stores will cut and sand one for you for about ten to fifteen dollars, depending upon its size. A 12 in.-wide magic mirror is perfect for evocations, so you should try to get a glass disc of this diameter. Make sure the piece of glass is free of imperfections before it is cut.

Hold your glass disc under running water and visualize any impurities in the glass being washed away. Then carefully towel dry it and lay it on some open newspapers or paper towels. Using a can of flat black spray paint, spray a light, even coat of paint onto one side of the mirror. Wait for it to dry, and then spray another coat. Repeat this until you can no longer see through your glass disc. Three or four coats should suffice.

If you have a display stand that is used for large plates, you can use this to hold your magic mirror as well. Try asking for one of these stands at a few department stores, if you don't have one around the house.

If you prefer, you can make your own stand. Cut a square of plywood that is about two inches wider than your mirror (14 x 14 inches if your mirror is twelve inches in diameter) and paint it black. Cut two triangles of the dimensions given in figure 4.15 out of plywood and attach them to the back of the plywood square as shown in figure 4.15, using brackets and wood screws. Make sure the wood screws do not go completely through the plywood or they will scratch your mirror.

To attach your mirror, center it on the front of your stand, making sure the painted side of the glass is facing the wood. Take six mirror mounts (you can find them in any hardware store) and screw them in at equally distant points around the mirror (see figure 4.15).

Your completed mirror and stand should be covered in a piece of black or white silk until consecrated. The consecration rituals for your magic mirror and other magical implements are given in chapter 5.

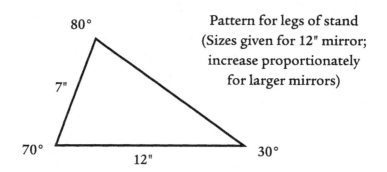

80°

7"

70°

12"

30°

Pattern for legs of stand
(Sizes given for 12" mirror;
increase proportionately
for larger mirrors)

FRONT OF STAND   Mirror mounts

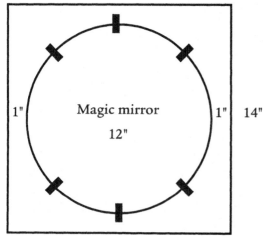

1"   Magic mirror   1"   14"
               12"

14"

BACK OF STAND

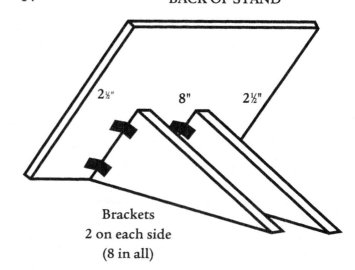

2½"   8"   2½"

Brackets
2 on each side
(8 in all)

FIGURE 4.15

chapter five

CONSECRATION RITUALS

I beseech thee to bestow upon this
weapon thy strength and fiery steadfastness, that with it I may control the spirits of thy realm for all just and righteous
purposes.

In chapter 4 we discussed the physical preparation of the implements used in magical
evocation. These tools cannot be used for
magic, however, until they are charged and
consecrated to the service of the Light. This
chapter contains the rituals and procedures
that should be performed to accomplish this.
To ensure your tools are properly dedicated to
the service of the Light, treat the following rites
with respect, remembering that the consecration of a magical instrument is a life-giving act.

Since the consecration of a magical object
is an act of will, it is best performed using
your Magic Wand and Sword. For this reason,

the first ritual given below is used to consecrate these instruments of the magician's will. Do not consecrate both the Sword and Wand at the same time, however, as the personal energizing of the first tool will take a lot out of you, if done properly, and will make it difficult to focus your energies on the second one. For best results, consecrate these tools on different days, or at least a few hours apart from each other.

When consecrating your Magic Wand or Sword, remember that you are accomplishing two things: dedication of the implement to the service of the Light, and charging of the implement with your own energy so it may effectively channel your will. The rest of the consecrations given in this chapter deal with tools that are either dedicated to Divine service or both dedicated to the Light and infused with some specific magical energy (for example, the Elemental Weapons).

The rituals given for the consecration of the Magic Wand and Sword and Elemental Weapons contain an abbreviated form of the Opening by Watchtower. Since your Elemental Weapons will not be ready for use until these consecration rituals are finished, they should not be used to perform a ritual opening yet. You should still, however, set up the altar with the Elemental Weapons in their respective quarters so they will begin to absorb the energies present there.

Consecration Ritual for Magic Wand or Sword

1. Prepare your altar and temple in your usual fashion (including Elemental Tablets on walls and Tablet of Union and tools on altar). For the simple ritual-opening segment, you will also need a red or white candle (which you should light before starting the ritual) and a cup full of water (not the Water Cup). Your Magic Wand or Sword (depending upon which you are consecrating) should be on your altar.

2. Perform step 2 of the Opening by Watchtower (see chapter 3).

3. Perform step 3 of the Opening by Watchtower.

4. Pick up the cup full of water with your left hand and, walking clockwise around your temple, sprinkle water along the perimeter of your magic circle with your right hand, while saying, *So therefore first, the priest who governeth the works of Fire must sprinkle with the lustral water of the loud resounding sea.*

5. Return to your position west of the altar, facing east, and place the cup of water back on the altar. Pick up the candle with your right hand and walk around your temple clockwise, retracing your magic circle in the air with the candle, and saying, *And when, after all the phantoms are banished, thou shalt see that Holy and Formless Fire, that Fire which darts and flashes through the hidden depths of the Universe, hear thou the Voice of Fire.*

6. Place the candle back on the altar, and perform step 15 of the Opening by Watchtower.

7. Perform step 16 of the Opening by Watchtower.

8. Perform step 17 of the Opening by Watchtower. This concludes the simple ritual opening.

9. Perform the Analysis of the Keyword, but instead of letting the Divine Light descend upon you, visualize the Divine Light as a sphere of white light descending upon your Magic Wand or Sword (again, depending upon which one you are consecrating). See the implement glowing in this Divine brilliance.

10. As you contemplate the instrument's glow, hold your hands above it, palms facing down, as if blessing it, and say the following, making sure to vibrate the name in bold capital letters: *Most Holy* **EHEIEH** (eh-hey-yay), *bless and*

consecrate this Wand (or Sword) that it may obtain the neces-
sary virtue, through Thee, Most Holy EHEIEH (eh-hey-yay),
Whose Kingdom endureth unto the Ages of the Ages.

11. Pick up the implement and hold it above your head with two hands, as if offering it to Heaven. Look up at the instrument and beyond it while saying, *To Your service do I dedicate this Wand (or Sword), Vast and Mighty One, in the hope You will find it to be an acceptable instrument with which I may aspire to the Light.*

12. Lower your Wand (or Sword) and, still holding it with two hands, perform the Middle Pillar Ritual up to step 8 (see chapter 3). As explained in the last part of step 8 in the Middle Pillar, draw the energy from the dark sphere at your feet up to your shoulders with an inhalation, and force the energy down your arms and into your implement with your next exhalation. Your exhalation should last long enough for you to mentally state the following: *I charge thee, creature of wood (or steel), to be an instrument of my will.* Feel this energy being absorbed by your Wand (or Sword). Repeat this channeling of energy six or seven times, and with each exhalation, mentally repeat the given affirmation.

13. Visualize an aura of white light surrounding your body and Wand (or Sword). Your implement has now become one with your magical being. Spend a few moments contemplating your new connection with your Wand (or Sword). It is now an extension of your body and an instrument of your will. Try to feel with its tip (or blade), as if it were an arm. After about three minutes of meditation in this fashion, your tool's consecration is complete.

14. Perform step 18 of the Opening by Watchtower.

15. Perform the Closing by Watchtower (steps 19–22).

16. Wrap your Wand (or Sword) in white or black silk and let no one else touch it.

Consecration of Elemental Weapons

Once your Magic Wand and Sword are consecrated, you can use them to perform the Golden Dawn consecration rituals for the Elemental Weapons. Some of the sigils and Hebrew names you painted upon these weapons will have to be traced in the air as you vibrate them. Check chapter 4 to make sure you are tracing the correct ones, as they are only written in Hebrew on your Elemental Weapons. I'm sure you noticed I did not explain the significance of these names in chapter 4. I did this because they are explained in the following rituals.

Do not consecrate more than one Elemental Weapon at a time. Even though you do not channel personal energy into these implements as you did with the Wand and Sword, the process of consecrating them is fairly lengthy and can cause fatigue if repeated in one period of magical working. You should be both physically and mentally relaxed when performing a consecration, so it is best to consecrate only one of these weapons per day, if possible. If you wish to get two done in a day, try performing the consecrations at least four hours apart from each other. Three consecrations in one day are definitely too many.

Consecration Ritual for Fire Wand

1. Prepare your altar and temple as usual. All the elemental weapons should be present on the altar, along with the consecrated Wand and Sword, the Tablet of Union, a cup full of water, and a red or white candle, which should be lit before the ritual begins. The altar and temple should be prepared in this fashion for the rest of the Elemental Weapon consecrations.

2. Perform the simple ritual opening, which is found in steps 2–8 of the Consecration Ritual for Magic Wand or Sword.

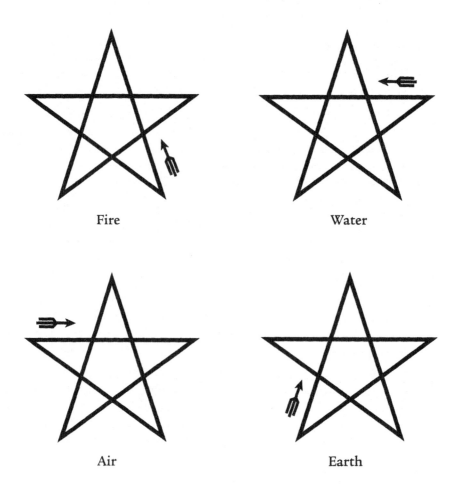

FIGURE 5.1

3. Pick up your Magic Wand in your right hand and your Fire Wand in your left and move clockwise to the south of your circle. Using your Magic Wand, trace a large blue Invoking Pentagram of Fire in the air over your Fire Wand.

4. Use your Magic Wand to trace the sigils and Hebrew letters of the names in capital letters given below in the air over your Fire Wand as you vibrate them (see chapter 4). Say, *O Thou, Who art from everlasting, Thou Who hast created all things, and doth clothe Thyself with the Forces of Nature as with a garment, by Thy Holy and Divine Name* ELOHIM (el-oh-heem) *whereby Thou art known in that quarter we name* DAROM (dah-rohm), *I beseech thee to grant unto me strength and insight for my search after the Hidden Light and Wisdom.*

5. Continue the oration, making sure to trace in the air above your Fire Wand the sigil and Hebrew letters of each name given in capital letters as you vibrate it. Say, *I entreat Thee to cause Thy Wonderful Archangel* MICHAEL (mee-chai-el) *who governeth the works of Fire to guide me in the Pathway; and furthermore to direct Thine Angel* ARAL (ah-rahl) *to watch over my footsteps therein.*

6. Continue the oration as follows: *May the ruler of Fire, the powerful Prince* SERAPH (seh-rahph), *by the gracious permission of the Infinite Supreme, increase and strengthen the hidden forces and occult virtues of this Wand so that I may be enabled with it to perform aright those magical operations for which it has been fashioned. For which purpose I now perform this mystic rite of Consecration in the Divine Presence of* ELOHIM (el-oh-heem).

7. Return your Magic Wand to the altar, pick up the Magic Sword in your right hand, and return to the south. Again, trace a large blue Invoking Pentagram of Fire in the air above your Fire Wand using your Magic Sword.

8. Make sure to vibrate the Enochian names, which are in capital letters, as you say the following: *In the Three Great Secret Holy Names of God borne upon the Banners of the South* OIP TEAA PEDOCE (oh-ee-pay tay-ah-ah peh-do-kay), *I summon Thee, Thou Great King of the South* EDEL PERNAA (eh-dehl pehr-nah-ah), *to attend upon this ceremony and by Thy presence increase its effect, whereby I do now consecrate this Magical Wand. Confer upon it the utmost occult might and virtue of which Thou mayest judge it to be capable in all works of the nature of Fire, so that in it I may find a strong defense and a powerful weapon with which to rule and direct the Spirits of the Elements.*

9. Say, *Ye Mighty Princes of the Southern Quadrangle, I invoke you who are known to me by the honorable title, and position of rank, of Seniors. Hear my petition, and be this day present with me. Bestow upon this Wand the strength and purity whereof ye are Masters in the Elemental Forces which ye control; that its outward and material form may remain a true symbol of the inward and spiritual force.*

10. Move clockwise around your circle once and return to the south. Hold up your Fire Wand and, making sure to vibrate the name in capital letters, say, *O Thou Mighty Angel* BZIZA (bay-zod-ee-zod-ah), *who art Ruler and President over the Four Angels of the Fiery Lesser Quadrangle of Fire, I beseech thee to bestow upon this weapon thy strength and fiery steadfastness that with it I may control the spirits of thy realm for all just and righteous purposes.*

11. Move clockwise to the west of your circle. Hold up your Fire Wand and say, *O Thou Mighty Angel* BANAA (bay-ahn-ah-ah), *Ruler and President over the Four Angels of Fluid Fire, I beseech thee to bestow upon this weapon thy strength and fiery steadfastness that with it I may control the spirits of thy realm for all just and righteous purposes.*

12. Move clockwise to the east of your circle. Hold up your Fire Wand and, remembering to vibrate the Angel name, say, *O Thou Mighty Angel BDOPA* (bay-doe-pay-ah), *Ruler and President over the Four Angels and Governors of the subtle and aspiring Etheric Fire, I beseech thee to bestow upon this weapon thy strength and fiery steadfastness that with it I may control the spirits of thy realm for all just and righteous purposes.*

13. Move clockwise around your circle to the north. Hold your Fire Wand high and say, *O Thou Mighty Angel* **BPSAC** (bay-pay-zah-cah), *who art Ruler and President over the Four Angels of the denser Fire of Earth, I beseech thee to bestow upon this weapon thy strength and fiery steadfastness that with it I may control the spirits of thy realm for all just and righteous purposes.*

14. Perform step 18 of the Opening by Watchtower.

15. Perform the Closing by Watchtower, with two differences. First, when you do the LBRP, use the Fire Wand instead of the dagger and, second, trace Banishing Pentagrams of Fire (see figure 5.1) instead of the normal banishing pentagrams, to make sure you properly banish the elemental forces invoked in this ritual.

16. Wrap your Fire Wand in its red, black, or white silk and let no one else touch it.

Consecration Ritual for Water Cup

1. Prepare your altar and temple as usual.

2. Perform the simple ritual opening, which is found in steps 2–8 of the Consecration Ritual for Magic Wand or Sword.

3. Pick up your Magic Wand in your right hand and your
   Water Cup in your left and move clockwise to the west of
   your circle. Using your Magic Wand, trace a large blue
   Invoking Pentagram of Water in the air over your Cup.

4. Use your Magic Wand to trace the sigils and Hebrew letters of the names in capital letters given below in the air
   over your Water Cup as you vibrate them (again, see chapter 4). Say, *O Thou, Who art from everlasting, Thou Who
   hast created all things, and doth clothe Thyself with the
   Forces of Nature as with a garment, by Thy Holy and Divine
   Name* EL (el) *whereby Thou art known in that quarter we
   name* MEARAB (mee-ah-rahb), *I beseech thee to grant
   unto me strength and insight for my search after the Hidden
   Light and Wisdom.*

5. Continue the oration, making sure to trace in the air
   above your Cup the sigil and Hebrew letters of each
   name given in capital letters as you vibrate it. Say, *I
   entreat Thee to cause Thy Wonderful Archangel* GABRIEL
   (gah-bree-el) *who governeth the works of Water to guide me
   in the Pathway; and furthermore to direct Thine Angel*
   TALIAHAD (tah-lee-ah-hahd) *to watch over my footsteps
   therein.*

6. Continue the oration as follows, *May the ruler of Water,
   the powerful Prince* THARSIS (thar-cease), *by the gracious
   permission of the Infinite Supreme, increase and strengthen
   the hidden forces and occult virtues of this Cup so that I may
   be enabled with it to perform aright those magical operations
   for which it has been fashioned. For which purpose I now perform this mystic rite of Consecration in the Divine Presence
   of* EL (el).

7. Return your Magic Wand to the altar, pick up the Magic
   Sword in your right hand, and return to the west. Again,
   trace a large blue Invoking Pentagram of Water in the air
   above your Cup, using your Magic Sword.

8. Make sure to vibrate the Enochian names, which are in capital letters, as you say the following: *In the Three Great Secret Holy Names of God borne upon the Banners of the West* EMPEH ARSEL GAIOL (em-peh-heh ahr-sell gah-ee-ohl), *I summon Thee, Thou Great King of the West* RA AGIOSEL (eh-rah ah-jee-oh-sell), *to attend upon this ceremony and by Thy presence increase its effect, whereby I do now consecrate this Magical Cup. Confer upon it the utmost occult might and virtue of which Thou mayest judge it to be capable in all works of the nature of Water, so that in it I may find a strong defense and a powerful weapon with which to rule and direct the Spirits of the Elements.*

9. Say, *Ye Mighty Princes of the Western Quadrangle, I invoke you who are known to me by the honorable title, and position of rank, of Seniors. Hear my petition, and be this day present with me. Bestow upon this Cup the strength and purity whereof ye are Masters in the Elemental Forces which ye control; that its outward and material form may remain a true symbol of the inward and spiritual force.*

10. Move clockwise to the south of your circle. Hold up your Water Cup and say the following, making sure to vibrate the Angel name: *O Thou Powerful Angel* HNLRX (heh-nu-el-rex), *Thou who are Lord and Ruler over the Fiery Waters, I beseech thee to endue this Cup with the Magic Powers of which thou art Lord, that I may with its aid direct the spirits who serve thee in purity and singleness of aim.*

11. Move to the west of your circle, going clockwise, and raise your Cup, saying, *O Thou Powerful Angel* HTDIM (heh-tah-dee-mah), *Thou who are Lord and Ruler over the pure and fluid Element of Water, I beseech thee to endue this Cup with the Magic Powers of which thou art Lord, that I may with its aid direct the spirits who serve thee in purity and singleness of aim.*

12. Walk clockwise to the east of your circle, raise your Cup, and say, *O Thou Powerful Angel* **HTAAD** (heh-tah-ah-dah), *Thou who are Lord and Ruler of the Etheric and Airy qualities of Water, I beseech thee to endue this Cup with the Magic Powers of which thou art Lord, that I may with its aid direct the spirits who serve thee in purity and singleness of aim.*

13. Move clockwise around your circle to the north. Hold your Cup high and say, *O Thou Powerful Angel* **HMAGL** (heh-mah-gee-ehl), *Thou who are Lord and Ruler of the more dense and solid qualities of Water, I beseech thee to endue this Cup with the Magic Powers of which thou art Lord, that I may with its aid direct the spirits who serve thee in purity and singleness of aim.*

14. Perform step 18 of the Opening by Watchtower.

15. Perform the Closing by Watchtower, with two differences. First, when you do the LBRP, use the Water Cup instead of the dagger, and, second, trace Banishing Pentagrams of Water (see figure 5.1) instead of the normal banishing pentagrams.

16. Wrap your Water Cup in its blue, black, or white silk, and let no one else touch it.

Consecration Ritual for Air Dagger

1. Prepare your altar and temple as usual.

2. Perform the simple ritual opening, which is found in steps 2–8 of the Consecration Ritual for Magic Wand or Sword.

3. Pick up your Magic Wand in your right hand and your Air Dagger in your left and move clockwise to the east of

your circle. Using your Magic Wand, trace a large blue Invoking Pentagram of Air over your Dagger.

4. Use your Magic Wand to trace the sigils and Hebrew letters of the names in capital letters given below in the air over your Air Dagger as you vibrate them. Say, *O Thou, Who art from everlasting, Thou Who hast created all things, and doth clothe Thyself with the Forces of Nature as with a garment, by Thy Holy and Divine Name* YOD HEH VAV HEH (yode-heh-vahv-heh) *whereby Thou art known in that quarter we name* MIZRACH (mitz-rach), *I beseech thee to grant unto me strength and insight for my search after the Hidden Light and Wisdom.*

5. Continue the oration, making sure to trace in the air above your Dagger the sigil and Hebrew letters of each name given in capital letters as you vibrate it. Say, *I entreat Thee to cause Thy Wonderful Archangel* RAPHAEL (rah-fay-el) *who governeth the works of Air to guide me in the Pathway; and furthermore to direct Thine Angel* CHASSAN (chah-sahn) *to watch over my footsteps therein.*

6. Continue the oration as follows: *May the ruler of Air, the powerful Prince* ARIEL (ah-ree-ehl), *by the gracious permission of the Infinite Supreme, increase and strengthen the hidden forces and occult virtues of this Dagger so that I may be enabled with it to perform aright those magical operations for which it has been fashioned. For which purpose I now perform this mystic rite of Consecration in the Divine Presence of* YOD HEH VAV HEH (yode-heh-vahv-heh).

7. Return your Magic Wand to the altar, pick up the Magic Sword in your right hand, and return to the east. Again, trace a large blue Invoking Pentagram of Air above your Dagger using your Magic Sword.

8. Make sure to vibrate the Enochian names, which are in capital letters, as you say the following: *In the Three Great Secret Holy Names of God borne upon the Banners of the East* ORO IBAH AOZPI (oh-row eh-bah-hah ah-oh-zohd-pee), *I summon Thee, Thou Great King of the East* BATAIVAH (bah-tah-ee-vah-hah), *to attend upon this ceremony and by Thy presence increase its effect, whereby I do now consecrate this Magical Dagger. Confer upon it the utmost occult might and virtue of which Thou mayest judge it to be capable in all works of the nature of Air, so that in it I may find a strong defense and a powerful weapon with which to rule and direct the Spirits of the Elements.*

9. Say, *Ye Mighty Princes of the Eastern Quadrangle, I invoke you who are known to me by the honorable title, and position of rank, of Seniors. Hear my petition, and be this day present with me. Bestow upon this Dagger the strength and purity whereof ye are Masters in the Elemental Forces which ye control; that its outward and material form may remain a true symbol of the inward and spiritual force.*

10. Move clockwise to the south of your circle. Hold up your Air Dagger and say the following, making sure to vibrate the Angel name: *O Thou Resplendent Angel* EXGSD (ex-jaz-dah), *thou who governest the Fiery Realms of Air, I conjure thee to confer upon this Dagger thy Mysterious Powers, that by its aid I may control the spirits who serve thee for such purposes as be pure and upright.*

11. Walk to the west of your circle, moving clockwise, and hold your Dagger, saying, *O Thou Resplendent Angel* EYTPA (eh-iht-pohd-ah), *thou who governest the Realms of Fluid Air, I conjure thee to confer upon this Dagger thy Mysterious Powers, that by its aid I may control the spirits who serve thee for such purposes as be pure and upright.*

12. Move clockwise to the east of your circle, raise your Dagger and say, *O Thou Resplendent Angel* **ERZLA** (eh-rah-zod-lah), *thou who rulest the Realms of Pure and Permeating Air, I conjure thee to confer upon this Dagger thy Mysterious Powers, that by its aid I may control the spirits who serve thee for such purposes as be pure and upright.*

13. Move clockwise around your circle to the north. Hold your Dagger up high and say, *O Thou Resplendent Angel* **ETNBR** (eht-en-bah-rah), *thou who rulest the Denser Realms of Air symbolized by the Lesser Angle of Earth, I conjure thee to confer upon this Dagger thy Mysterious Powers, that by its aid I may control the spirits who serve thee for such purposes as be pure and upright.*

14. Perform step 18 of the Opening by Watchtower.

15. Perform the Closing by Watchtower, with two differences. First when you do the LBRP, use the Air Dagger, instead of the dagger you normally use. Second, trace Banishing Pentagrams of Air (see figure 5.1) instead of the normal banishing pentagrams.

16. Wrap your Air Dagger in its yellow, black, or white silk, and let no one else touch it.

Consecration Ritual for Earth Pentacle

1. Prepare your altar and temple as usual.

2. Perform the simple ritual opening, which is found in steps 2–8 of the Consecration Ritual for Magic Wand or Sword.

3. Pick up your Magic Wand in your right hand and your Earth Pentacle in your left and move clockwise to the north of your circle. Using your Magic Wand, trace a large blue Invoking Pentagram of Earth over your Pentacle.

4. Use your Magic Wand to trace the sigils and Hebrew let-
    ters of the names in capital letters given below in the air
    over your Earth Pentacle as you vibrate them. Say, *O Thou,*
    Who art from everlasting, Thou Who hast created all things,
    and doth clothe Thyself with the Forces of Nature as with a gar-
    ment, by *Thy Holy and Divine Name* ADONAI (ah-doe-nye)
    whereby Thou art known in that quarter we name TZA-
    PHON (tsa-phon), *I beseech thee to grant unto me strength*
    and insight for my search after the Hidden Light and Wisdom.

5. Continue the oration, making sure to trace in the air
    above your Pentacle the sigil and Hebrew letters of each
    name given in capital letters as you vibrate it. Say, *I*
    entreat Thee to cause Thy Wonderful Archangel AURIEL
    (ohr-ee-el) who governeth the works of Earth to guide me in
    the Pathway; and furthermore to direct Thine Angel PHOR-
    LAKH (phor-lahk) *to watch over my footsteps therein.*

6. Continue the oration as follows: *May the ruler of Earth,*
    the powerful Prince KERUB (keh-rub), *by the gracious per-*
    mission of the Infinite Supreme, increase and strengthen the
    hidden forces and occult virtues of this Pentacle so that I may
    be enabled with it to perform aright those magical operations
    for which it has been fashioned. For which purpose I now per-
    form this mystic rite of Consecration in the Divine Presence of
    ADONAI (ah-doe-nye).

7. Return your Magic Wand to the altar, pick up the Magic
    Sword in your right hand, and return to the north.
    Again, trace a large blue Invoking Pentagram of Earth
    above your Pentacle using your Magic Sword.

8. Make sure to vibrate the Enochian names, which are in
    capital letters, as you say the following: *In the Three Great*
    Secret Holy Names of God borne upon the Banners of the
    North EMOR DIAL HECTEGA (ee-mohr dee-ahl hec-
    tey-gah), *I summon Thee, Thou Great King of the North* IC
    ZOD HEH CHAL (ee-kah zohd-ah heh kah-la), *to attend*
    upon this ceremony and by Thy presence increase its effect,

whereby I do now consecrate this Magical Pentacle. Confer upon it the utmost occult might and virtue of which Thou mayest judge it to be capable in all works of the nature of Earth, so that in it I may find a strong defense and a powerful weapon with which to rule and direct the Spirits of the Elements.

9. Say, *Ye Mighty Princes of the Northern Quadrangle, I invoke you who are known to me by the honorable title, and position of rank, of Seniors. Hear my petition, and be this day present with me. Bestow upon this Pentacle the strength and purity whereof ye are Masters in the Elemental Forces which ye control; that its outward and material form may remain a true symbol of the inward and spiritual force.*

10. Move clockwise to the south of your circle. Hold up your Earth Pentacle and say the following, making sure to vibrate the Angel name: *O Thou Glorious Angel* **NAAOM** (nah-ah-oh-em), *thou who governest the Fiery essences of Earth, I invoke thee to bestow upon this Pentacle the Magic Powers of which thou art Sovereign, that by its help I may govern the spirits of whom thou art Lord, in all seriousness and steadfastness.*

11. Walk to the west of your circle, moving clockwise. Hold up your Pentacle and say, *O Thou Glorious Angel* **NPHRA** (ehn-phrah), *thou who governest the moist and fluid essences of Earth, I invoke thee to bestow upon this Pentacle the Magic Powers of which thou art Sovereign, that by its help I may govern the spirits of whom thou art Lord, in all seriousness and steadfastness.*

12. Move clockwise to the east of the circle, raise your Pentacle and say, *O Thou Glorious Angel* **NBOZA** (ehn-boh-zohd-ah), *thou who governest the Airy and Delicate Essence of Earth, I invoke thee to bestow upon this Pentacle the Magic Powers of which thou art Sovereign, that by its help I may govern the spirits of whom thou art Lord, in all seriousness and steadfastness.*

13. Walk around your circle, clockwise, to the north. Hold your Pentacle high and say, *O Thou Glorious Angel* **NR-OAM** (ehn-roh-ah-ehm), *thou who governest the dense and solid Earth, I invoke thee to bestow upon this Pentacle the Magic Powers of which thou art Sovereign, that by its help I may govern the spirits of whom thou art Lord, in all seriousness and steadfastness.*

14. Perform step 18 of the Opening by Watchtower.

15. Perform the Closing by Watchtower, with two differences. First, when you do the LBRP, use the Earth Pentacle instead of the dagger, and second, trace Banishing Pentagrams of Earth, which are the normal banishing pentagrams you trace in the LBRP.

16. Wrap your Earth Pentacle in its black or white silk, and let no one else touch it.

Consecration Ritual for Implements of the Art

With your Wand, Sword, and Elemental Weapons consecrated, you can then perform the Opening by Watchtower, instead of the simple ritual opening, when you consecrate the rest of your implements. The Consecration Ritual for Implements of the Art can be used to consecrate all your other magical tools except the Tablet of Union and Elemental Tablets, which are consecrated by the performance of the Watchtower Ritual, and the Triangle of the Art, which is consecrated at the beginning of each evocation (more on the Triangle later on in this book).

Unlike the previous tools, there is an effective way to consecrate the rest of your magical implements (such as your mirror and censer) at the same time. The only limitation here is how many tools you can fit on your altar at one time. Remember, your Wand, Sword (or Dagger), and Elemental Weapons have to also be on the altar. Therefore, because of space limitations, you may have to perform two separate consecration rituals to dedicate all your tools.

1. Set up your altar and temple as usual. Place the tools to be consecrated on top of the Tablet of Union in the center of your altar.

2. Perform the Opening by Watchtower, up to and including step 17.

3. Perform the Analysis of the Keyword and draw the sphere of white brilliance above you onto the unconsecrated tools on your altar. See your implements glowing in this white light.

4. As you contemplate the glow of the instruments on your altar, hold your hands above them, palms facing down as if blessing them, and say the following, making sure to vibrate the name in capital letters: *Most Holy* **EHEIEH** (eh-hey-yay), *bless and consecrate these instruments of the magical art, that they may obtain the necessary virtues, through Thee, Most Holy* **EHEIEH** (eh-hey-yay), *Whose Kingdom endureth unto the Ages of the Ages.*

5. Pick up one implement and hold it above your head with two hands, as if offering it to Heaven. Look up at the instrument and beyond it, while saying, *To Your service do I dedicate this (name of instrument), Vast and Mighty One, in the hope You will find it to be an acceptable instrument with which I may aspire to the Light.*

6. Repeat step 5 for each of the implements on your altar.

7. Perform step 18 of the Watchtower Ritual.

8. Perform the Closing by Watchtower.

9. Wrap each instrument in either a black or white piece of silk and let no one else touch them.

Having mastered the techniques, rituals, and exercises given in the book so far, and having prepared your magical tools, you can consider yourself a magician who is ready to learn the art of magical evocation. The rest of the book deals with the theories of why and how evocation works and the techniques for making it work. Remember, the work you have done so far has completely prepared you for success in this magical practice, and you should approach the art with confidence in yourself and your abilities.

From this point on, your life will never be the same again. The changes you will create and observe in the world around you will seem truly miraculous at first, but as your understanding of the occult universe increases, you will begin to see just how natural these manifestations of your will actually are.

chapter six

MAGICAL THEORY

In a way, sigils are like psychic transmitters that can help a magician send messages to a spirit.

The applied sciences of the world have provided us with some incredible technological advances, all of which are based upon some type of scientific expertise used in a practical application. For example, a telescope works because its design is based upon the laws of refraction and reflection; a particle accelerator increases the velocity of electrons and protons to near the speed of light thanks to our understanding of electromagnetic laws; and people everywhere can change TV channels without getting out of their seats thanks to our knowledge of encoding infrared light. By applying the natural sciences (including physics and chemistry) to the solution of everyday problems, we have been able to make our lives much easier. Like physics and chemistry, magic is also a science that can be applied to this end.

The practice of magical evocation can be thought of as an applied science because its application depends upon knowledge of the science of magic. Other magical applied sciences include talismanic magic, candle burning, herbalism, and knot magic. Even though these practices are all inherently different, one still has to know how magic works to use them effectively.

If you think back to the beginning of the book, you will remember the differences between armchair theorists and practicing magicians. An armchair theorist is a person who reads a lot of books and does nothing with the knowledge he or she acquires. A practicing magician, on the other hand, is an applied scientist, similar to an engineer, who uses known principles to accomplish practical goals. The goal of this book is to help you become one of the latter. After all, learning how magic works and using it to perform evocations will enable you to create wonderful changes in your life.

Therefore, to make sure you are ready to move on to the actual techniques of this magical science, this chapter will teach you the occult principles at work in magical evocations. This includes explanations of how conjurations work, how sigils help magicians contact entities, and why these beings can be commanded when evoked. The theories given in this chapter provide a solid base of knowledge that can be applied to the evocation techniques given later.

Magic

Magic is the most powerful science in the world. When its principles are applied successfully, this science enables a magician to make changes in the universe in accordance to his or her will. But unlike other sciences, magic can affect other planes of existence, as well as the farthest reaches of the physical universe. This multiplanar facet of magic must be understood before one can hope to be effective at working rituals. For this reason, what follows is a look at the nature of the three planes, followed by a discussion on how magical rituals work.

As I've already mentioned, there are three planes on which magic works in the universe. They are the physical, astral, and mental planes. As humans, we are the only beings on Earth who inhabit all three of these planes of existence simultaneously in three "bodies"—our physical bodies, souls, and spirits. This is possible because the planes are not separated like rungs on a ladder. Instead, they coexist all around us. Visiting one of the planes, therefore, does not require traveling, but rather entering the right subtle body within our physical body to experience the plane it occupies.

The physical plane is where our material bodies reside. Whatever can be experienced by our physical senses directly, or through some type of scientific detection device, also resides on this plane. There is no need to go into detail to describe this plane as it is simply the material world of which everyone is aware since birth. Both time and space bind the physical plane in the magical universe, making it possible to measure both accurately on this plane.

The astral plane is where our astral bodies or souls reside. To experience the astral plane fully and directly, the astral senses must be developed or trained (as taught in chapter 2). Sometimes called the formative plane, the astral plane is where ideas begin to take form. Magic performed on this plane has the powerful advantage of being brought one step closer to materialization, as any form built on the astral will eventually appear in some way on the physical plane.

Since all three planes coexist, a magician who travels astrally (as introduced in chapter 2) will find he or she can visit any place in the physical world by viewing its astral counterpart. Due to the fact that these counterparts sometimes change from their astral stages of development to their eventual physical ones, however, impressions gained through astral travel may seem a little distorted at times. Astral traveling is not the best way to see the physical world.

When thinking of the astral plane, keep in mind that it is a more subtle plane of existence. Not everyone is consciously aware of its existence, mostly because it doesn't fit into the physical, rule-based world science has defined over the years. The astral plane is bound to the universe only by space, which makes it possible to travel through time in this plane.

The last plane we'll be dealing with is the mental plane. This is where our spirits or "minds" reside. Sometimes called the creative plane, the mental plane is where pure ideas are created and from where we often receive inspiration. Bound by neither time nor space, the mental plane serves two important magical functions: It allows us to contact any mind or intelligence from any time or place in the universe, and it enables us to create ideas that can take form on the astral plane and eventually manifest in the physical world. The first of these two functions will be dealt with in chapter 8 in the discussion on contacting entities on the mental plane, while the second is dealt with below.

Having explained the existence of planes in the universe, we can now apply this knowledge to the explanation of how magical rituals work. To help illustrate the basics, let's take a close look at the following Hermetic axiom: "To know, to will, to do, and to be silent." Within these ten words, the most basic occult principles are found.

1. **To Know:** Before any magic can be performed, you have to know and understand a specific magical technique. If you wish to perform an evocation, you have to know, step-by-step, how to perform an evocation ritual, and you must understand what is occurring in each part of the ceremony.

2. **To Will:** The power that makes a magical ritual work comes from the magician's will. To work any type of magic, you have to have a strong desire to attain some end, and then be able to channel that desire in a ritual. An evocation is usually performed with two desires kept in mind: the desire that the called entity will appear, and the desire that it will accomplish the task which the magician will assign.

3. **To Do:** The ritual has to be performed correctly for the desired effect to be achieved. A magician has to do more than go through the motions and pronounce the words of a rite, he or she also has to be able to supply the necessary power and perform the necessary mental workings (such as visualizations and affirmations).

4. **To Be Silent:** While few people actually practice this, keeping silent about what type of magical work you are doing is crucial to a successful outcome. By telling everyone that you just did a ritual for so and so reason, you are allowing them to build up unconscious thought-forms that could be destructive to your ritual's success. It is best to not only keep quiet about what type of magic you are performing, but also about the fact you are performing magic at all.

These four principles must be followed closely if you wish to succeed in magic. It is not enough to know that magic is the science of causing change through the use of will; you also have to realize that magic, like every other science, follows some kind of system. If you spend some time meditating on the four points just brought up, you will probably find the answer to several questions you had regarding the practice of magic.

The banishing rituals given earlier teach one how to raise and control psychic energy, which can be used to empower a thought-form. If one were to simply visualize his or her goal, empower it, and release it, it would manifest on the mental plane, astral plane, and eventually the physical plane. This is the basis for all magical workings, regardless of type. As we shall soon see, this simple form of magic can also become destructive to a ritual's success if there are doubts of a ritual's outcome present in a magician's mind.

In a magical evocation, your calling of the entity is done on the mental plane. After it "hears" you, it either comes to the astral or physical planes, depending upon the type of evocation you are performing. The calling of the entity is performed on the mental plane because all magic begins in the mind, is powered by the will, and causes change.

Why do evocations work? Why do entities feel compelled to come to the magician when called? To answer this question, I want to refer to something I mentioned at the beginning of the book. When a magician stands in the center of the circle, he or she is able to invoke the power of Divine Providence. In the Opening by Watchtower, a vortex of power descends upon the magician, which the magician can use to empower the ritual he or she is performing. Since this power comes from God, who is invoked in the Watchtower Ritual, the magician can in effect command Holy Energy, granting him or her Divine Authority.

When the magician who is in touch with the Godhead in this fashion calls out to an entity, that being has no choice but to answer. The only limiting factor in this power of the magician is knowledge. The magician has to have a very good idea of the nature of the entity he or she is calling and the plane it comes from. When these facts are known, the magician is able to direct his or her granted Divine Authority toward commanding a specific being.

Speaking to an entity with this type of authority will only work if the magician believes in the power he or she is channeling. Actually, you have to do more than just believe in it, you have to become it. A magician performing an evocation has to radiate authority. The voice used has to be confident and steady, and even posture has to be carefully controlled. You can't call an entity to the physical plane with a soft, insecure voice, and with your head hunched forward. You have to stand up straight and let your command be heard throughout the farthest reaches of all the planes in the universe (of course, I don't mean you should scream at the top of your lungs, as this may slightly upset your neighbors, particularly at night). When performing evocations to the astral plane, however, the type of voice used is different (see chapter 7) to ensure the process of skrying isn't disrupted.

Know that the entity hears you and that it has to appear. Believing the entity will manifest is important. Confidence in your abilities ensures success, because you can devote all your energy to the ritual. By doubting the evocation will work, you are accomplishing little more than destructive magic. Consider the following example: You have reached the peak of your ritual. The power in the room is so intense that you feel your body pulsing with it. With your Magic Wand in your right hand, you prepare to call out to the entity. All of a sudden, however, instead of pronouncing the call, you think to yourself, "Is this really going to work?" Remember what I said earlier about magic beginning in the mind? This thought, pronounced at the absolute worst time in the ritual, can use up a good amount of your raised energy to create a thought-form named: "Is this really going to work?" This thought-form will then circle you as the evocation proceeds, causing your doubts to grow. Its presence will be felt as you keep thinking about the possibility of your ritual's failure. Soon, you will not even be paying attention to what you are really trying to accomplish.

The thought-form created in the above example is only one of the reasons why an evocation can fail if you doubt it will succeed. The authority granted to you when you perform an evocation makes it appear to the entity that God is calling it. If you doubt an evocation will work, then you are not letting this Divine Power flow through you properly. You have to become an almost mythic being to perform an evocation, because you are attempting to control what most people would consider mythic forces. Accepting the ability to act with Divine Influence is not easy for most people. It requires a delicate balance of humility and authority. You must consider yourself below God, and at the same time, equally powerful. In other words, you should act confident since God is helping you perform the evocation, and at the same time humbly give thanks for this aid.

Conjurations

At the heart of any evocation is a verbal calling to the spirit commonly known as the conjuration. You don't have to look hard to find an example of one of these, as virtually every grimoire contains one or more. The *Goetia* contains several different ones that should be recited in the order they are given, in case the entity decides not to show up.

Why is it necessary to repeat conjurations? After all, if a magician calls an entity with the proper authority, shouldn't it come after one calling? Actually, if the magician is able to establish a link with an entity, it should appear after one conjuration. Each of the two types of evocation, however, contain different methods for contacting the entity and require different uses of conjurations.

In evocation to the astral plane, the magician has to establish initial contact with an entity using its sigil (more on sigils later). If the technique used to accomplish this is performed successfully, then there will only have to be one oration given to the spirit (which must be memorized, as explained in chapter 7). When performing evocations to the physical plane, however, the process of contacting the entity actually takes place after the conjuration is given, eliminating the need for repeating it. If the other parts of an evocation are performed properly, there only has to be one conjuration given to the spirit.

Now that we've established the need for only one conjuration, you might ask yourself, "Which one should I use?" According to some grimoires, if you do not pronounce the orations word for word, the entity will either not show up at all or will appear before the magician angry and attempt to exact some type of wrath. Both of these warnings are absolutely ridiculous. They are only included in books to make a dabbler think twice about using the rituals of High Magic. I rarely use conjurations from grimoires, and I have yet to be carried off by an angry demon! And even if the legends were true, it is impossible for an evoked entity to cross the boundary of the magic circle. Of course, only a magician would know that, making these threats quite intimidating to the uninitiated.

Ignore any threats like these if you come across them. If you want to use a conjuration given in a book, you should only use it if you feel it helps you get the appropriate feel for the ritual. Sometimes, conjurations contain bizarre-sounding words of power that help a magician think he or she is using some type of sonic key in a ritual. This same effect is also obtained by using conjurations that are written in another language. If you feel this effect helps you perform an evocation, then by all means use one of these types of orations. I do recommend, however, that you try to find out what it is you are saying if the oration is entirely in another language. It's always good to know what commands you are giving so you can enforce them with your will.

Now we come to a shocking and little-known fact about conjurations: You could write your own! There is no reason why a conjuration you write can't work as well as one you find in a book, as long as you follow a few simple rules. Conjurations accomplish two important functions in a ritual, and you have to make sure to include statements that accomplish them. These are the two functions conjurations perform: They verbally establish the magician's authority over the entity, and command the entity to appear. While most grimoires contain unique conjurations, they still accomplish these two functions, regardless of how they are worded. Let's look at an excerpt of a conjuration from the *Goetia* to illustrate some of the elements that accomplish the two afore-mentioned functions ("N." is substituted for the name of the spirit being evoked):

I do invocate and conjure thee, O Spirit N.; and being with power armed from the Supreme Majesty, I do strongly command thee, by Beralanensis, Baldachiensis, Paumachia, and Apologiae Sedes; by the most Powerful Princes, Genii, Liachidae, and Ministers of the Tartarean Abode; and by the Chief Prince of the Seat of Apologia in the Ninth Legion, I do invoke thee, and by invocating conjure thee. . . . Also by the names Adonai, El, Elohim . . . I do exorcise thee and do powerfully command thee, O thou spirit N., that thou dost forthwith appear unto me here before this circle in a fair human shape, without any deformity or tortuosity . . .

I chose these lines from the first conjuration in the *Goetia* to help illustrate some of the most common elements of the orations found in grimoires. First of all, note that the first few lines of the conjura-tion establish the magician's Divine Authority over all the mentioned spirits ("and being with power armed from the Supreme Majesty"). As you can see in the example, following this affirmation of authority, spirits are called in the names of both benevolent and malevolent enti-ties. In this case, the names are significant because they are of the spir-its who rule over the entities found in the *Goetia*, and for this reason the infernal names included in conjurations are different in each gri-moire. This part of the oration is not usually necessary, as the real power in an evocation comes from the Most High. On occasion, how-ever, a spirit does have a ruler who might be detaining it for some rea-son (see chapter 1), and this ruler's name should be known so it can be commanded to let its servant appear. Either way, Divine Authority overrules any other influences affecting the entity being called.

The conjuration goes on to give some of the Names of God, which should be vibrated in a ritual for best effect (in the rituals of evocation given later, names to be vibrated are given in bold capital letters, as in the other rituals in this book). This part of the oration reinforces the magician's fortitude, by once again invoking Divine Names and establishing his or her link with the Godhead. When reciting a conjuration, a magician should try to feel the energy he or she has accumulated with each vibration of a Divine Name, to help strengthen his or her resolve.

Another common element of a conjuration is the part where the magician commands the entity to appear before him or her in a "fair human shape, without any deformity or tortuosity." The wording of this command varies from grimoire to grimoire, but its general purpose is clear: The spirit is told to appear in a form that is not disturbing to behold. When dealing with angels, elementaries, and planetary intelligences, this is not a problem. The nature of these entities is such that their natural forms are not at all terrible to view. Demons, on the other hand, find it extremely difficult, if not impossible, to comply with this wish of the magician. They are grotesque beings by nature and take on appearances that reflects this. The above-mentioned command in a conjuration will by no means work wonders on infernal entities, and if you choose to work with demons, do not expect them to assume the appearance of beautiful humans, or for that matter, any remotely pleasing form. Again, because of their highly deceptive and wicked nature, I strongly suggest you avoid working with demonic entities.

Remember, to write your own conjuration, you have to include a segment that reestablishes your Divine Authority over the entity you are calling and actually commands the entity to appear before you. That's it! There is no need to include pages of curses, demanding that the entity appear or else; you don't have to look up a million different Divine Names to include in your oration (not that you'd find that many anyway). Simply include a few God Names that you understand and in the power of those names command the entity to appear. Some conjurations also contain explanations of each name given, with mythological or Biblical references to the power of the name. You could certainly include these, if you feel they make your conjuration stronger. Of course, make sure what you include is correct. Here is an excerpt of one such oration, taken from *The Greater Key of Solomon:*

I conjure ye by the name Tetragrammaton Elohim, which expresseth and signifieth the Grandeur of so lofty a Majesty, that Noah having pronounced it, saved himself, and protected himself with his whole household from the Waters of the Deluge.

The best way to include such a reference is to either copy it from a grimoire, or to look up the Kabbalistic significance of one of the Divine Names of God in a suitable reference (Regardie's *Golden Dawn* and Crowley's *777* are both excellent) and include a statement of the name's meaning.

The final option you have in writing your own conjuration is to include a statement that relates to your religion. Every religion has its own names for God, and in some cases, Goddess. A true magician should feel free to substitute names and mythological tales that correspond to his or her beliefs. In the following excerpt of a conjuration taken from the *Grimoire of Honorius* (which was supposedly written by Pope Honorius of the Catholic Church, although this is questionable), elements of Christianity are brought into the oration:

I, X. [the magician's name], do conjure thee, O Spirit N. [the spirit's name], by the living God, by the true God, by the holy and all-ruling God, Who created from nothingness the heaven, the earth, the sea and all things that are therein, in virtue of the Most Holy Sacrament of the Eucharist, in the name of Jesus Christ, and by the power of this same Almighty Son of God, Who for us and for our redemption was crucified, suffered death and was buried; Who rose again on the third day and is now seated on the right hand of the Creator of the whole world, from whence He will come to judge the living and the dead.

Parts of this conjuration were adapted from the Roman Catholic Mass, to help the magician, who is in this case Christian, relate to the Divine forces he or she is invoking. The same can be done with any religious ceremony or text. An invocation in a Witch's Book of Shadows can be used, as can a passage from the Koran. The examples of conjurations given later in this book adhere to Kabbalistic names and references, but as a magician, you should feel free to change these conjurations to suit your own beliefs.

While we are on the subject of changing God Names in rituals, it is important to note the following: The Golden Dawn rituals given earlier in the book contain Hebrew names which, through their repeated use, have become words of power that are critical to the success of these rituals. Changing these rituals is not recommended as most of their potency will be lost. If you do not feel comfortable with Kabbalistic references and are a practicing magician of another tradition, then your best alternative is to use whatever opening rituals you normally use within your current tradition. The same goes for the Four Elemental Weapons, as you may want to substitute whatever tools are used in your opening. You should be able to adapt the structure of the evocation rituals given in this book to any ceremonial magic tradition you are quite familiar with.

Another way to increase the potency of a conjuration is to add a statement about the entity's nature after its name. For example, *I conjure thee, O Spirit Phalegh, Olympian Ruler of the sphere and province of Mars. . . .* Including this type of statement helps maintain a stronger contact with the entity while you are performing the conjuration, as you are constantly reminded of its origin and nature. Carefully word your identification of the entity so it is short and precise. You don't want to ramble on and on about the identity of the spirit. Simply stating its name, office, and sphere or realm of origin is fine.

Now that you know the basic elements of a conjuration, you can feel free to write your own. You might want to check the basic structure of the conjurations given in the next two chapters, though, to get an idea of how all the pieces fit together. Once you write one you feel comfortable with, you should try to memorize it. This will be easy, because you will be using the same conjuration for every evocation you perform. Of course, your oration will have to be modified slightly when performing the two different types of evocations. For evocations to the astral plane, you will command the spirit to appear in the crystal or mirror. In evocations to the physical plane, you will command it to appear in the triangle before you. Again, use the conjurations given later in the book as models for how to word this.

Of course, you don't have to write your own conjurations. You can either use one you find in a grimoire or one in this book. If you do decide to write your own, however, you should treat it as a magical project. Sit in a consecrated circle as you write and try to think of the potency of the words you are putting onto paper, and what they mean to you. When you are done with your conjuration, copy your final draft on to the first page of a new notebook, which you will be using to record the results of your evocations. Consecrate this book using the rituals in chapter 5, as you will want to have it in the circle with you when you perform evocations.

The notebook just introduced will be used for more than just recording the results of your rituals, and reading your conjuration from. In this notebook, you will also copy the sigil of each spirit you work with and its description (see chapter 9). The reason for copying the attributes and descriptions of the entity is so you can verify the entity's identity through questioning (see chapter 1). When you evoke an entity, you can sketch in your book what it looks like to you and write down any information it gives you regarding future evocations of that entity. Some spirits will give you special conjurations or words of calling to use, which will ensure they come quickly when called in the future. It is to your advantage to write these down in your book and to use them.

Another important use for this notebook is one I mentioned in chapter 1. You can have the entity sign its name in the book to verify its identity (in evocations to the physical plane only), by sliding your opened book into the triangle using your sword (never let your arm or any other body part pass over the boundary of the circle). The name will appear on the paper as an astral impression you can behold, using your trained astral vision. This is similar to the way Edward Kelly was able to view the letters of the Enochian Alphabet for the first time. They appeared in a pale yellow color on paper so he could trace them.

The signatures/impressions you receive will appear in different colors, depending upon the nature of the spirit, and will fade away by the end of the ritual. For this reason, you might want to trace a spirit's signature into your book when you first behold it if you want a permanent record of what it looks like. Whether or not you do so, however, isn't really important. What is important is that you are able to read the entity's name, because experience has shown that if a signature is legible and spelled correctly the evoked spirit is being honest about its identity.

Grimoires

When working with grimoires, you will find that each has its own magical system or mythological background. For example, the *Goetia* contains the names of spirits that were supposedly once trapped into a vessel by King Solomon; and the *Necronomicon* (the version edited by Simon) contains, among other things, the sigils of a legion of fifty spirits that were assigned to the Mesopotamian god Marduk. The differences between the grimoires of the world has often made magicians wonder if any of them are based on factual occurrences.

If you think back to chapter 1, you'll remember our discussion concerning entities and grimoires that were probably made up by armchair theorists with nothing better to do. Do not let the authenticity of grimoires worry you; after all, if they have been in use by magicians, they will work. Period. It doesn't matter if the *Key of Solomon* was written a thousand years after Solomon's death. Whether the entities already existed or were simply made up, their existence today cannot be disputed. So if you find a grimoire that you are skeptical about, consider its first date of publication, its notoriety, and its subject matter. If it is a well-known tome and has been around for at least a decade or so, then it should most likely work.

The key factor will then be its subject matter. For example, if you find a book that deals strictly with fictional mythologies (like the *Necronomicon* edited by George Hay, see below), then there is a good chance its entities will be extremely difficult to evoke (because you will probably have to create them yourself, like you would an egregore). To illustrate the reasons behind this, I will use the example of the two grimoires with the same name, both of which were inspired by the fiction of a famous early twentieth-century writer.

Howard Phillips Lovecraft (1890–1937) is recognized today as one of the most important horror writers of all time. His tales of the "Cthulhu Mythos," which only achieved real fame after Lovecraft's death, dealt with the premise that a race of Ancient Ones, who ruled the Earth hundreds of millions of years ago, lurks just outside our perception, behind some kind of mystical gates, awaiting their return as rightful rulers of the Earth. The time for this conquest of the Earth was identified as the time when "the stars would be right." In these tales, an ancient tome of magic known as the *Necronomicon* was used to summon some of these beings to visible appearance. The fame of this nonexistent book of magic spread for decades after Lovecraft's death, until two different versions of the book "surfaced."

The first *Necronomicon* to hit the stands in our nonfictional universe was edited by a mysterious individual known only as Simon and released in hardcover form on the winter solstice in 1977. It has since been reprinted several times in paperback form by Avon Books and has achieved an underground following of sorts. Even necklaces of some of the sigils in the book can be found in occult bookstores today. As a final note on the popularity of this tome, the original hardcover is sold as a collector's item today for five hundred dollars and up!

Some of the entities found in Lovecraft's stories are also found in this grimoire, although they are spelled differently (i.e., Lovecraft's "Cthulhu" became "Kutulu"). What is interesting, however, is the way they are used. The magical system in the book is based on actual Mesopotamian mythology and the entities from the Cthulhu Mythos are included as demonic beings in adaptations of famous Sumerian myths, such as "Inanna's Descent into the Underworld" and the Enuma Elish, or Sumerian Creation Myth. Most of the rituals in the book deal with astrally visiting the Planetary Spheres and their rulers, who correspond to the actual deities associated with the planets in Mesopotamia.

Is this book a fake? Even though it contains several rituals that are similar to the type of magic practiced in Sumer (i.e., knot magic), and Sumerian words, the book is probably a well-researched hoax (like several other grimoires). However, its rituals work well because they are, to some extent, based upon actual magical practices. If you want to experiment with this book, I especially recommend you try evoking a few of Marduk's Legion of Fifty. These spirits are mentioned in the last part of the Enuma Elish and most of them are very helpful.

The second *Necronomicon* was released in Great Britain in 1978 and was edited by George Hay. Even though it includes a lengthy essay on how it was decoded from a cipher manuscript attributed to John Dee, one cannot ignore how similar the rituals and chants in it are to the ones in the works of H. P. Lovecraft. In fact, the rituals found in this book are nothing more than a good example of plagiarism, and the book has been proven to be a fabrication. There is nothing in this grimoire even remotely based on actual occult practices, and I can't imagine why someone would attempt to work the rituals found in this *Necronomicon*.

The purpose of the previous comparison should be clear. Any grimoire you find could be a fake, but it doesn't mean you can't use one that fits into some kind of occult system. A good way to spot a useful grimoire is to ask yourself the following questions: Is it based on some kind of mythology or magical belief system you know exists? Does it contain instructions for evocations that resemble those in other grimoires (this includes mention of a crystal, mirror, or triangle; use of a magic circle and other types of known magical implements)? Does it contain spirit origins you are familiar with (such as planetary and elementary)? Has it been around for a long time? Is it a translated work, and if so, where is the original manuscript located?

Worrying about which grimoires to work with is not necessary at this point. The spirits listed in chapter 9 should keep you busy for quite some time. But when you decide to look for more spirits to summon, you will want to turn to traditional tomes of magic (see the bibliography at the end of this book). When you acquire a few of these books, however, don't worry about the instructions given in them for performing evocations. The only rituals you will ever need to evoke entities are given in this book. Grimoires will actually only give you two basic pieces of information you should find useful: the names and descriptions of spirits and their sigils, the latter of which deserve an indepth explanation.

An entity's sigil or seal is a symbol that helps the magician establish a link with that entity. In a way, sigils are like psychic transmitters that can help a magician send messages to a spirit. The two types of sigils you will be working with are traditional and derived. Traditional sigils are ones that have been in use for hundreds of years, if not longer. They are found in grimoires, and it isn't really clear how they were created or discovered. Some were undoubtedly created through the instinct of ancient magicians, while some were either viewed clairvoyantly or given to magicians by the spirits themselves. If a grimoire contains a sigil for an entity, then you should have no problem contacting that being; simply use the sigil as described in chapters 7 and 8. But what do you do when you have a name of an entity you want to summon and no sigil to do it with? The answer is simple: make your own.

There are actually several types of spirits who do not have traditional sigils. A good example of these are the various angels found in the Sephiroth of the Tree of Life. To devise sigils for each of these angels, or for any entity, you can use one of two versions of a glyph known as the Rose Cross (see figure 6.1). The original Rose Cross Lamen of the Golden Dawn contains Hebrew letters, and you should use this glyph if you know the Hebrew spelling of an entity's name (usually an angel). To create sigils for entities with non-Hebrew names, I have included a version of the glyph with Roman characters.

Note: Some of the petals on the Rose contain more than one letter, as the Hebrew alphabet has fewer letters, which still make up all of the sounds in the Roman alphabet.

Creating sigils with either of these glyphs is actually quite easy. To begin, lay a piece of thin paper over the Rose Cross and draw a little circle over the first letter in the entity's name. Then draw a line going from the circle to the rose petal containing the second letter in the entity's name. Continue by making a line that connects the end of your first line with the third letter in the entity's name. When you come to the last letter in the name, make a little slash. (See figure 6.2 for an example of this process, using Roman characters and the name of an imaginary spirit, Fribo.) If two or more sequential letters lay in the path of a straight line, then you will have to make a little loop in the line to indicate where each of these are (see figure 6.2). You may also have to use the same petal twice if a name has a double letter or two letters that are represented by the same petal. If this happens, make a double hump in the sigil to show you are using that petal twice in a row.

In the listing of spirits in chapter 9, I have used the Hebrew Rose Cross to make sigils for angels. This version must be used with these beings as the spelling of their names in Roman characters adds extra letters, making the sigil different. See figure 6.3 for an example of the difference between the sigil of Raphael done correctly in Hebrew letters and incorrectly in Roman characters. When dealing with Hebrew names, always use the Hebrew version of the Rose Cross as Hebrew spellings of names have power in themselves.

With this method, you can create very effective sigils for any being you find in a grimoire, and even a being you create yourself. In chapter 10 you will find instructions for creating your own egregores, and this Rose Cross method can be used to prepare the sigils of these beings once you name them.

Now that some of the basics of evocation are covered, we can move on to the actual practice itself. Chapter 7 contains complete instructions for successfully evoking entities to the astral plane. Make sure you try this form of evocation first! Evocations to the physical plane are much easier to perform once you have succeeded at astral ones. If you have been practicing the exercises and rituals in the book so far, you are already prepared to work the magical techniques found in the next chapter. Familiarize yourself with the process of astral evocation as soon as possible. This way you can pick a goal you would like to accomplish magically, find an entity in chapter 9 which is described as being able to fulfill your wishes, and evoke it. By doing this, you will be able to prove to yourself that evocation works, and by achieving results you will be able to call yourself a true practicing magician.

Hebrew Characters

Roman Characters

FIGURE 6.1

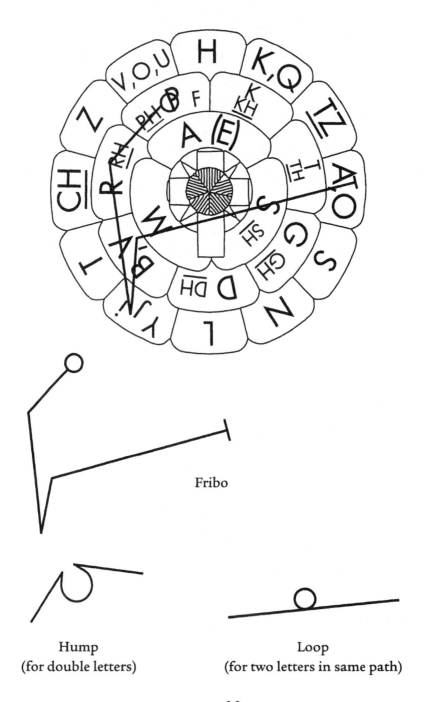

Fribo

Hump
(for double letters)

Loop
(for two letters in same path)

FIGURE **6.2**

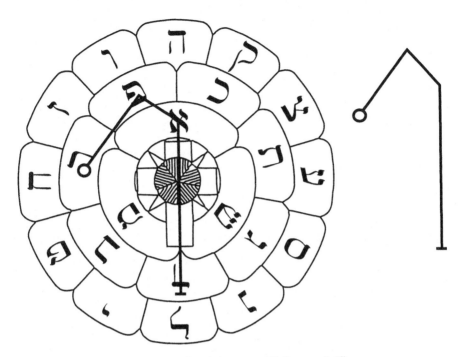

Raphael (correct Hebrew sigil)

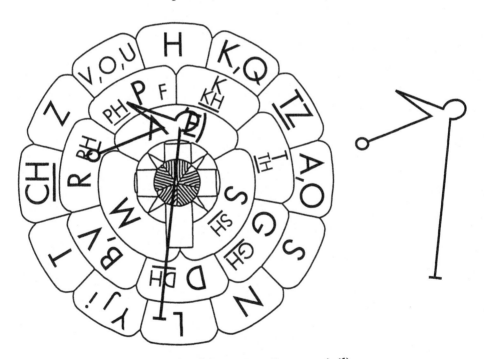

Raphael (incorrect Roman sigil)

FIGURE 6.3

chapter seven

EVOCATION TO
THE ASTRAL PLANE

> When your astral vision awakens, you
> will begin to perceive the shape of the
> summoned entity forming within your
> crystal or mirror.

E ven though evocation to the astral plane
is the easier of the two techniques of this
magical art, it should not be viewed as a begin-
ner's exercise. While it is true that astral evocation yields excellent results for the beginning
magician, and that it is a prerequisite to evoking entities to the physical plane, you will find
yourself using this technique even after you
have had some success at the latter. In fact,
there are some major advantages to calling an
entity only to the astral plane and viewing it
through some apparatus such as a magic mirror or crystal. The following benefits of using

astral evocations are included as an introduction to the many uses and facets of this practice.

When an entity is brought to the astral plane, it is not subject to any limitations regarding how it may appear or what it may show to the magician. For example, let's suppose you want an entity to help you find something you have lost. In this case, an evocation to the physical plane would not be as effective as one to the astral because the entity called in the former matter would have to try to explain the location of said object, while the entity appearing in the mirror or crystal can actually "show" you its location by creating a visual image of it in the mirror.

Remember, all manifestations begin on the mental plane and pass through the astral and physical planes. An entity only has to will something to appear on the astral and it will, but if an entity on the physical plane tried to do this it would be exceedingly difficult, as the vibrations of this plane would never completely be in accordance with the being. As explained in chapter 8, to physically evoke an entity you have to try to make the temple room correspond closely to the being's sphere of origin. This can never be completely successful on the physical plane, because there are too many influences which are nonconcordant with many types of entities. For example, the presence of physical air in a room does not agree with a Fire's elementary nature. These hindrances to an entity's nature are not present on the astral plane. Only etheric matter exists there, and entities have much greater control over this type of matter than they do over physical types, because their spheres of origin are also etheric. Of all the entities in the universe, the ones that can affect the greatest direct control over the physical plane are the elementaries, who can control their element's physical representation (see chapter 1); but as I just mentioned, even their powers are diminished by the presence of other elements.

Let me make something clear before we move on. Because of their control over the astral plane, most entities are less constricted when evoked to this plane. Do not let this statement mislead you, however. All entities can affect changes on the physical plane, regardless of the type of evocation used. The limiting factor we are discussing here is the amount of time needed. If you need an entity to "show" you something, then you will have to evoke it astrally, but if you want an entity to help bring something into your life (such as money or love), you can use either method with equal success, as it will take from a few hours to a few weeks for your goal to manifest, regardless of how the spirit was evoked.

Another reason why calling an entity to the astral plane is advantageous is the ease at which it can be accomplished. Because of the simple physical preparations necessary for the temple room and the uncomplicated nature of the technique itself, evocation to the astral plane is a good choice for a magician who is in a hurry or who doesn't have a lot of time to work. As we will discuss in chapter 8, physical evocations take a long preparation time, because the room has to be turned into a representation of the entity's sphere of origin. This is not at all the case with astral evocations.

The temple used in astral evocations is prepared in basically the same way as it was for the other rituals in this book, with a few additions. Set up your altar as usual, with all the magical tools in their respective places, and burn an appropriate incense according to the nature of the entity (see correspondences in chapter 8). If you cannot get an incense that corresponds to the entity being summoned, then you might want to try an equal mixture of frankincense and gum mastic, which is a very "evocative" incense I often use. On the east side of the altar, you should place a chair so when you sit in it you are facing east. This direction can be used for all types of entities except elementaries. For these beings, use the direction that corresponds to their element.

Outside the circle, hang the Elemental Tablets on the walls as usual. In the quarter you will be facing when seated, position a table about three feet away from the circle. Upon it, place the Triangle of Art with the apex of the triangle facing away from the circle. In the center of the circle within the Triangle, place your magic mirror or crystal so that when you are seated you can look into it. Use a table that elevates your skrying medium to a comfortable height. A table as tall as your altar should be about right.

The final consideration in your temple's preparation is that of lighting. The room should be dark, except for the light of a lamp or two candles on the altar. Make sure you can't see the reflection of your light source in the crystal or mirror when you are seated. Use whatever method you used when you were practicing skrying to accomplish this. I use an oil lamp positioned directly behind me on the altar, which works very well.

One of the most important benefits of the art of evocation to the astral plane is the speed at which the summoned spirits appear. Since they only need to build up a form on the astral plane, entities find it easy to appear in forms true to their nature without difficulty. When a being tries to appear on the physical plane, it has to build a "body" for itself out of material that is not completely in accordance with its nature. Since this is not a problem on the etheric astral plane, entities can appear quicker.

The above reasons should give you an idea why you may decide to practice evocation to the astral plane even after you succeed at the more difficult conjuration of entities to the physical plane. Of course, for now, the most practical reason for working with this technique is obvious. It must be mastered before moving on to physical evocations.

The Four Elements of Astral Evocation

If you have been practicing the techniques given in the book so far, then you will find astral evocations to be incredibly easy to master. The basic concepts and procedures of this technique can be reduced to four, easy-to-understand elements: (1) the calling of the spirit; (2) the viewing of and communication with the spirit; (3) the commanding of the spirit; and (4) the dismissal of the spirit. Each of these elements must be understood before evocations are attempted.

The calling of the spirit is accomplished in two steps in an astral evocation. After the magician performs the Opening by Watchtower, he or she has to sit in the chair before the altar, facing the mirror or crystal. Still holding the Magic Wand in the right hand, the magician must hold the sigil of the entity in his or her left hand and look at it. The sigil should be drawn in an appropriate color ink on a new piece of white paper. Using the same gaze as if skrying, the sigil should be stared at for a few minutes while the name of the entity is repeated mentally. This should be repeated for as long as it normally takes for the magician's skrying faculties to awaken. When this amount of time has elapsed, the first stage of contact is complete. Using the sigil as a psychic transmitter, the magician has now established a link with the entity.

After a link between the magician and the entity is formed, it is then possible to command the entity to appear. To do this, the magician must look up from the sigil in his or her left hand and gaze into the mirror or crystal. While doing so, the previously memorized conjuration to the spirit should be pronounced as the magician begins skrying. This conjuration has to be spoken firmly and with Divine Authority, but should at the same time be pronounced in a low and somewhat monotonous voice to help maintain concentration on the process of skrying. The entity should manifest on the astral plane in a matter of moments, and the magician should become aware of its presence in the normal amount of time it takes for his or her skrying faculties to awaken.

Here we come to the second important part of an astral evocation: the viewing of and communication with the entity. Once the entity appears within the crystal or mirror, the magician has to establish its identity. As explained earlier, a good knowledge of the entity's nature, coupled with a list of suitable correspondences in the magician's book, should help in this matter. Asking an entity to sign its name in the space in the mirror or crystal is a good way of verifying an evoked spirit's identity. Once the magician is certain he or she is dealing with the desired being, an address of welcome should be made (an example of this is given in the "Ritual for Evocation to the Astral Plane" section later in this chapter). From this point, basic communication with the entity is possible. Keep in mind that speaking in a low and monotonous voice, as discussed above, helps maintain concentration on skrying and should be employed whenever communication occurs with an entity. Some entities may wish to give the magician secret rituals or words of power which he or she can use to summon the entity in the future. If this information is given at all, it will most likely occur at this point in the evocation, and as with every detail of an evocation, it should be recorded in the magician's book.

After some basic communication (there should not be an extended period of pointless communication, only topics relating to the purpose of the evocation should be discussed), it is time to perform the third important part of an evocation to the astral plane: the commanding of the spirit. Once the general subject matter of the magician's objective has been discussed, the magician must clearly and concisely give a command to the evoked spirit. This should be carefully thought out before an evocation, as the entity will follow the magician's instructions to the letter. Even if the purpose of an evocation is to obtain information, an oration should still be given to the spirit, clearly stating this. This way, the spirit will give very focused answers to a magician's questions regarding an area of its expertise. Here is an example of a command given to an evoked entity:

I command and constrain you, O Spirit N., by the power of Yhvh, the true unpronounceable name of God, to give me true and honest counsel in the arts of divination, that through these practices I may gain accurate insight into all occurrences, regardless of location or time.

When creating a command like this, the magician has to be sure to include a detailed but short description of what he or she wants the entity to do. There should always be a reference to the Divine Power with which the magician is commanding the entity, to ensure obedience. Also, if you request anything more than information from an entity, you have to include a clause that commands the entity to harm no one while performing its task. Let me explain this last point fully.

Most entities in the universe have little or no concept of the difference between good and evil. If they are commanded to do something, they will look for the easiest way to accomplish their goal. Let's say a magician asks a spirit to get her landlord "off her back" about being late with the rent. Think about what options this poorly worded command gives the spirit. On the very next morning, the landlord could be found dead in bed, apparently the victim of a heart attack, with a look of terror upon his face. This magician has indirectly put out a magical "hit" on her landlord and is now karmically responsible for his death. Even if she didn't want to have her etheric avenger kill her landlord, she still practiced a form of black magic by giving the entity a command that made this option available.

Avoiding a calamity like this in any type of magic is simple. Word your commands to entities, or even candle spells and talismanic charging rituals, so they include a clause about not having anyone come to harm as a result of your ritual.

After the magician is certain his or her wishes are understood by the entity, the next and last step in the evocation is the dismissal of the spirit. A magician should include in his or her dismissal command a brief statement of thanks to the entity for coming quickly, and a statement that orders the entity to return to its sphere of origin in peace. An example of this type of dismissal is given in the evocation ritual later in this chapter. Once the entity disappears from view, the magician can perform the Closing by Watchtower. The LBRP and BRH included in the Closing ensure that the entity is completely banished to its sphere.

These procedures make up the basic outline of an astral evocation ritual. As you can see, the process itself is rather simple. If performed by a trained magician, the results of astral evocations can be quite remarkable. Any spirit found in any grimoire can be conjured and commanded using this type of evocation, which means there is no real limit to the success a magician can achieve through this magical practice.

Keeping the Established Link

At the heart of any astral evocation is the process of using a crystal or mirror as a window to the astral plane. While performing evocations, you will begin to notice this "window" aspect for yourself as you find astrally evoked entities are not really present in the crystal or mirror, but in the nearby astral plane. When you skry, you can see and hear this entity in the same way you could see and hear someone talking to you through an open window. Can this link you establish with an entity be broken accidentally?

If you are performing evocations by yourself, you may have to look away from your mirror or crystal at some time to either look up some information or write down something the spirit said. It is okay to do so. Remember, the spirit will stay in the nearby astral plane until dismissed, so do not worry about looking away for a moment. After you return your gaze to the mirror or crystal, however, prepare yourself for the possibility of seeing nothing. This may happen if you divert your attention for too long, as your astral senses will naturally "deactivate." Simply gaze into your skrying medium again for a few moments, and you will perceive the image of the spirit coming into focus once more. It may take you a few seconds to reawaken your astral sight, but the entity itself will remain there, waiting to be seen, until you dismiss it.

Some magicians do not like astral evocations because they have a hard time sitting down to skry after raising power in the Watchtower Ritual. There is another alternative for people with this problem. Astral evocation is a process that is suitable for use by two magicians at the same time, as was the case with John Dee and Edward Kelly. In this case, one magician performs the evocation while the other does the actual skrying. For this to work, both magicians gaze at a sigil of the spirit for an agreed upon amount of time. After that, the magician who is standing begins the conjuration, while the seated magician stares into the crystal or mirror. For the length of the conjuration, the standing magician asks all the questions of the spirit and commands it, while the skryer does nothing but report what he or she sees.

This method works well, but it only gives a magician a partial feeling of success when the evocation is over. Unlike participation in an occult lodge, where a specific magical curriculum is followed, working with another person on a regular basis can actually hinder the development of magical abilities, unless a systematic rotation of tasks is followed. In other words, if you want to work with another person at performing evocations, then you should make sure to switch roles regularly, to ensure you both completely develop your magical abilities. Many occult biographers agree that Dee couldn't skry, and that he had to work with Kelly for this reason. This probably wasn't the case, however. I think Dee couldn't skry because he never worked at it, and by allowing Kelly to do all the skrying for him, he made sure he would never attain this faculty. If you work with another person, make sure you switch roles each time you perform an evocation.

Ritual for Evocation to the Astral Plane

Having looked at some of the benefits and facets of astral evocation, we can now move on to the practice itself. This section includes the complete ritual for evoking spirits to the astral plane. It is given in a form that is suitable for one magician to use; however, it can be adapted for use by two magicians by following the guidelines given above.

Read the ritual several times to make sure you understand all of its parts. Memorization of the conjuration given below or the one you wrote yourself is necessary, as you will have to recite it while skrying. The conjuration I have written for use in this ritual contains the Divine Names of God that are used in the four quarters in the LBRP, as these names are familiar to you by now. They are given in the order they are vibrated in the LBRP, so you should have no problem remembering them. Memorization of the conjuration will not present a problem if you perform this ritual with another magician, as the person reciting the conjuration can read it, since he or she will not be skrying.

1. Prepare your temple room as usual. Place chair, table, Triangle, and skrying medium (mirror or crystal) in their appropriate positions (see above). Place your book containing information about the entity under the chair so you can use it to verify the entity's identity during the ritual. Have altar lighting arranged so it is not visible in the skrying medium when seated and make sure you have ample incense appropriate to the working for your censer. The sigil of the entity to be evoked should be placed on the Tablet of Union and wrapped in a piece of black silk until it is needed. Before starting the ritual, put on your robe and magic ring if you have them.

2. Perform the Opening by Watchtower up to and including step 17 (see chapter 3).

3. Unwrap the sigil and take it into your left hand. Pick up your Magic Wand with your right and move clockwise to your chair.

    Note: If you are evoking an entity that you feel may not be obedient (i.e., a demon), use your Magic Sword instead of your Wand as its purpose is to enforce the magician's will. (Again, you should avoid calling demonic entities in the first place.)

4. Sit in your chair, take a slow, deep breath, and exhale. Repeat this until you feel relaxed and centered.

5. Look at the sigil in your left hand with a relaxed stare, as if you were skrying. While you do this, mentally repeat the name of the entity and imagine your astral call is echoing throughout the universe. Continue gazing at the sigil and repeating the name for about as long as it normally takes for your skrying vision to awaken.

6. Having established a psychic link with the entity, look up into the mirror or crystal on the table in front of you and allow yourself to skry. As you do this, recite your conjuration or the following: *I evoke and conjure thee, O Spirit N., by the power and authority of the Supreme Majesty—the true God Who is known by the names of* YOD HEH VAV HEH (yode-heh-vahv-heh), ADONAI (ah-doe-nye), EHEIEH (eh-hey-yay), *and* AGLA (ah-gah-lah)—*to appear before me within this mirror (or crystal), in a fair and comely shape.*

7. Continue to skry. When your astral vision awakens, you will begin to perceive the shape of the summoned entity forming within your crystal or mirror. When the image becomes clear, ask the entity, *What is your name?* If it answers correctly, move on to step 8. If it gives you a statement, check it with what you know about the entity. For example, an Elemental Ruler may say, "I am the King of Fire." If it gives you a correct answer of this nature, move on to step 8. If the entity claims to be another being, then

you can either ask the entity its purpose in coming to you or command it to return to its sphere of origin and repeat your evocation from step 5.

8. Ask the entity to sign its name in the space around it, within the mirror or crystal. If it is unable to do so or if it gives you an incorrect name, then you should command the entity to return to its sphere of origin and repeat the evocation from step 5. If the entity signs its name correctly, then you can proceed to step 9. *Note:* If you have to repeat the evocation and still contact the wrong entity, move on to the dismissal in step 11, perform your closing, and try again later. If you perform the evocation properly, this should rarely be necessary, as the correct entity will most likely come the first time.

9. Once you are convinced of the entity's identity, welcome it by saying, *In peace I welcome you, Spirit N., and in the name of the Most High I command you to remain within this mirror (or crystal) until you are dismissed, to speak honestly, and to truthfully answer all questions put before you.*

10. Speak to the entity concerning your purpose for evoking it. When you feel you have established its ability to accommodate your wishes, give it your carefully worded command (as explained earlier in this chapter).

11. When you feel the entity understands your wishes, and you have learned all you wish to know, give the entity the following command of dismissal: *Go in peace and return to your sphere of origin, O Spirit N. By the authority of the True God, I command you to harm none as you depart, and to be ready to come quickly if called again.*

12. Wait for the spirit to disappear from view and pick up your book. Write down any important information, including any rituals or words of power the entity may have given you.

13. Stand up and walk clockwise to your normal position behind the west of the altar, facing east. Wrap the sigil in black silk and place it on the Tablet of Union once again.

14. Perform step 18 of the Watchtower Ritual.

15. Perform the Closing by Watchtower.

After the ritual is completed, you should try to ground yourself with some type of no-magical activity. Then, after you feel you have returned to normal consciousness, make a complete entry into your book of the results of your operation. Do not confuse this entry with the one you should make in step 12 of the above ritual, as the recording of words of power and rituals should be made immediately if they are granted.

The sigil used in the evocation should not be used again. You could carry it around with you as a kind of talisman or you could burn it, effectively releasing its magical energy. If you decide to carry it around with you, however, make sure to burn it after your magical goal is obtained. The proper destruction of sigils is a necessary part of astral evocations, but as we will see in chapter 8, when entities are evoked to the physical plane, they can impart lasting occult virtues to more permanent sigils (such as ones engraved on metal), which you can keep to draw these influences into your life.

Applying Information from Entities

Once you have succeeded at your first evocation and you begin to practice the art on a regular basis, you will soon find yourself in possession of a great amount of recorded information. Since spirits will answer all questions regarding topics they are familiar with, and sometimes, unfortunately, topics they know nothing about, you may end up with books full of revelations, not all of which are to be trusted. How do you sort the good data from the bad?

First of all, make sure you do not immediately act on whatever information you get. Some spirits like to entertain their hosts by discussing subjects they are not masters of. A good way to prevent this from happening is to stick to subjects the entity is described as understanding. However, even if you do only ask pertinent questions, make sure to take the answers with a proverbial grain of salt. Don't burn down your new house just because an entity tells you it is full of evil vibrations. If you do feel uncomfortable in your new home, try performing banishing rituals in each room and purifying them with fire and water as in the simple opening ritual given in chapter 5. If an entity makes you aware of such presences, it may even give you a good way to banish the influences if you ask.

You should always be wary of unverifiable facts. If a spirit tells you that people will eat nothing but seaweed in ten years, don't sell off your possessions to raise capital for a processing plant. Check out all spirit-recommended business ventures thoroughly before starting them. Maybe seaweed will replace salad in a decade, but then again, maybe it won't. If the industry never boomed, what would you do with five-hundred-thousand pounds of processed seaweed?

While very few people would go to the extents mentioned above, these scenarios serve as examples of what could happen if you take a spirit's advice without thinking about it first. Don't let these examples discourage you, however. If you ask the right questions to the right entity, and carefully research the information you receive, you will find it to be accurate most of the time. Then, after the information is verified, you can use it to your advantage in everyday life. Medical doctors can discover new cures for diseases, archaeologists can learn the locations of hidden artifacts and treasures, lawyers can get advice on winning certain cases, and single adults can find out where to go on what night to meet their ideal mates. No matter who you are, and to what end you apply it, knowledge is extremely powerful.

Let me say one final thing about the information you receive from entities and how you apply it. You are responsible as a magician and as an ethical person for all your actions, whether they are magical or non-magical in nature. Under no circumstances should you let any harm come to another person as a result of your actions. If an entity tells you to kill your boss because he is planning to fire you, you should not only refrain from buying a handgun and using it, but you should also ask yourself if you are really dealing with a beneficial type of entity. If you find an entity is telling you to do things you find unethical, you may want to avoid working with that being in the future.

After a few successes at astral evocations, you may want to start working on the next technique: evocation to the physical plane. The two prerequisites to this practice are belief in the possibility of evocation, and skill in the technique of astral traveling. Successfully performing evocations to the astral plane takes care of the first prerequisite. As for the second, it was introduced in chapter 2, and if you have had some success with those preliminary exercises, you should find the more advanced ones in the next chapter fairly easy.

Along with the technique of astral traveling, the next chapter includes all the preparations required for evoking entities to the physical plane, and of course, a complete evocation ritual.

chapter eight

EVOCATION TO
THE PHYSICAL PLANE

When you behold the entity on the mental plane, you can then bring it back with you to the physical one.

In chapter 2, you read the preliminary exercises for accomplishing astral traveling. At the end of these exercises, you also found a method for traveling to the elemental regions using the Tattwas. If you have successfully accomplished astral traveling to these realms and have performed astral evocations as detailed in chapter 7, then you are ready to begin evoking elementary entities to the physical plane. If you haven't succeeded at this form of astral traveling, however, then perhaps this chapter will provide you with enough additional information to make the practice a little easier to master.

The purpose of teaching astral traveling in this book should be clear from the explanations given earlier. Evoking an entity to the physical plane instead of the astral requires extra effort on the part of the magician. Passing through the astral plane and manifesting on the physical one is not easy for a spirit to accomplish on its own. This is the reason why "ghosts" often look different to a number of observers in one room and are not seen at all by others. Most "ghosts" are entities that reside on the astral plane. Therefore, those who see them often have some form of natural astral vision, and those who see nothing have undeveloped astral senses. "Ghosts" rarely manifest on the physical plane.

Because it is difficult for a spirit to manifest physically on its own, the magician must "guide it through" the planes. It is not egotistical for us to think we have this ability as magicians, in fact, it is a proven occult fact. Think for a moment about what really happens when you practice astral traveling. By using an etheric body of light (an astral construct), you are able to move your perception away from your physical body and into that of your astral body. Once this is accomplished, you can then travel to another sphere using some kind of portal (i.e., Tattwa). These other spheres you travel to all exist on the mental plane. Therefore, the portals help you enter your spirit and access the correct region of the mental plane you wish to visit. When you want to return, you exit through the portal you entered from, effectively leaving the mental plane and your spirit, and become aware of your astral body or soul once again, fully entering the astral plane. Then you enter your awareness into your physical body to complete the process of returning to the physical plane.

The above process, which is exactly what is taught in this book, shows how magicians can move through the planes in either direction at will. In fact, moving from the mental plane to the physical one is actually easier for a magician than going in the other direction, because by returning to the physical plane we are actually returning to our own plane of origin. The magical universe, like the physical one, is a realm of equilibrium. If a rock is thrown into the air, gravity pulls it back to Earth to maintain that equilibrium. Likewise, if a magician leaves his or her own plane, then the universe will help that magician return. The same also holds true for entities.

Having accepted this magical law, you should realize that the following is also true. Since we are able to move through the planes if magically trained, and can return with little magical effort, our mystical energies can be used in other ways on these return trips from the mental plane. It is therefore possible to contact an entity on the mental plane and guide it to our own physical plane using the powers of concentration available to us, as a result of the ease with which we can come back. Likewise, when it is time to banish the entity, it can leave our physical plane and return to its own with equal ease. This is another example of the magical universe making sure equilibrium is maintained.

Once you understand the above discourse, evocations to the physical plane should seem much more feasible. If you've ever tried to "conjure a spirit" in the past (that is, pick up a grimoire and read out loud, as most magicians do when they first buy these books), you probably found yourself standing in a drawn circle, calling out to the air with a trembling voice, expecting a flash, some smoke, and a hooded creature to appear.

When these "conjurations" fail, most would-be magicians look at other magical methods and find that even the simplest spells or rituals require something more than a verbal or physical component. Every magical practice has its own "secret" that makes it unusable by those who do not understand it. Reading a spell over a lit candle will not bring you what you desire, but empowering the candle and visualizing your goal while you read the spell will. Drawing some symbols on a piece of paper will not protect you from illness, but magically charging that piece of paper as a health talisman will. Magical evocation to the physical plane is no exception, as it too has its secret: astral traveling. Without the use of this technique, physical evocations will not work.

Astral Visitation of the Planetary Spheres

Assuming you have been practicing the exercises given in chapter 2, there will come a time when you are able to astrally travel to the elemental regions through the Tattwas. As I mentioned earlier, you are ready to evoke elementary beings when you accomplish this. When that time comes, familiarize yourself with the ritual in this chapter, select an elementary spirit from chapter 9, and evoke it. Remember, practice is a very important part of becoming a successful magician. The following astral traveling technique will allow you to journey to planetary spheres and consequently to evoke the inhabitants of them.

Aside from the elementary spirits, whose realms you can visit through the Tattwas, there are two other classifications of spirits you may want to summon: entities that come from a planetary sphere or from the Kabbalistic Sephiroth, and entities that seem to come from no known sphere. To establish communication with entities of the first kind, it is necessary to use a technique that is in many ways similar to the one for visiting the elemental spheres. This technique is given below. Establishing contact with entities that do not come from known spheres requires a slightly different procedure, which we'll deal with later.

To travel to a particular Sephirah, many magicians use a technique known as path-working, where a magician travels from Malkuth (our Sephirah), along the paths on the Tree of Life to another Sephirah. This system tends to be rather tedious for the purposes of evocation, as traveling along the Tree in this manner can take quite awhile. For this reason, this technique is better used by magicians who would like to receive visions of the Sephirah, and not by those who wish to call the denizens of these spheres to the physical plane. Instead of giving this system here, I will explain the system I use for traveling to other spheres, which is much quicker and easier to use.

Before you can use this system, you have to first make a set of nine colored discs. These discs should be about four or five inches in diameter. You can cut them out of colored card stock or make them out of thin pieces of wood, which can be painted. The correct planetary symbol of each disc should be painted or drawn on both sides. Table 8.1 shows the colors of the discs, the planetary symbols that go on them, and the colors each symbol should be drawn in. Once these discs are ready, you can start practicing the following technique. Unlike the skrying of the Tattwas, you will not try to create a complementary image of the cards. Instead, you will have to remember what they look like, and recreate them in your mind's eye during the following procedure.

The exercise below is given in ritual format because it should be performed within a temple that has been purified through banishing rituals. For your early attempts at this exercise a chair can be used (see step 1), however, as soon as you succeed at astral traveling to planetary spheres, begin practicing standing up, as you will have to do so during an evocation to the physical plane.

1. Prepare your temple as usual. Have a chair positioned on the east side of your altar, facing east. You can lay the disc of the sphere you wish to visit on this chair, for now. Adjust your lamp so the room is dimly lit.

2. Perform the LBRP and the BRH.

3. Move clockwise to the chair and pick up the disc with your left hand.

4. Sit in the chair and look at the disc for a few moments. Close your eyes and try to imagine a sphere of glowing energy, the same color as the disc, floating above your head.

5. Using the technique you learned in chapter 2, pull this colored light into yourself to create an astral body of light.

6. Rise up out of your physical body in your colored body of light and move up toward the glowing sphere above you. Become aware of the planetary symbol that is inscribed upon it in the same color as the symbol on your disc. Notice how this sphere above you is a large three-dimensional representation of the smaller disc you made.

7. Allow yourself to rise higher, and pass through the glowing portal.

8.   Once you pass through the sphere, look around you. Try to notice any surroundings that stand out. If any beings are present, vibrate the Divine Name of the Sephirah given in the correspondence tables in table 8.2 (remember, the planetary spheres and the Sephiroth are the same). Only a beneficial being will be able to approach you once you do so.

9.   When you are ready to return, pass through the glowing sphere you entered through and allow yourself to return to your body.

10.   Open your eyes and when you feel ready, perform the LBRP and BRH.

When performing the above exercise, you create an astral body of light in the color that corresponds to the Sephirah you want to travel to. By doing so, you allow yourself to vibrate on the frequency of that sphere, making travel there easy. Mastering this technique should be simple if you have successfully performed the exercises given earlier in the book, as the basic process of leaving your body is the same.

Once you can perform this technique standing up, you are ready to summon planetary entities to the physical plane using the "Ritual for Evocation to the Physical Plane" given later in this chapter. You may have noticed by now that I seem to urge you to practice what you learn at every opportunity. The reason I have reiterated the importance of practice is because most of the rituals and exercises in this book can only be performed after the ones given before them are mastered. This is especially the case with the different types of evocation in this chapter. Before summoning entities with unclear spheres of origin, for example, you have to first master calling the spirits from realms you are familiar with (again, elemental or planetary).

Many grimoires contain spirits that clearly do not come from any known spheres. When you come across a listing of an entity of this type that you wish to summon to the physical plane, you will have to rely on a slightly different technique than the one you already learned. We will deal with this technique at the end of this chapter.

The Eight Elements of Physical Evocation

To illustrate how the process of astral traveling and the rest of the parts of a physical evocation ritual fit together, let's look at the following step-by-step description of the process. After the temple is prepared both physically and magically, these are the steps of a physical evocation: (1) The contemplation of a symbol representing the entity's sphere and/or its sigil; (2) the recitation of the conjuration; (3) the astral journey to the entity's sphere of origin and initial contact with the entity; (4) the bringing of the entity to the physical plane; (5) the verification of the being's identity; (6) the greeting of and communication with the entity; (7) the commanding of the spirit; and (8) the dismissal. Let's look at each of these steps below.

The first step of a physical evocation is one you are already familiar with in part. If the entity comes from either a planetary or elemental realm, then you should contemplate the correct disc or Tattwa, respectively. In addition, it is also necessary to contemplate the sigil of the entity and to repeat its name a few times to help establish a link. When it becomes time to actually travel astrally to the sphere, you will have to use these symbols again. Notice I state "and/or" in the paragraph above when referring to the use of the sigil. I did this because only the sigil of the entity is contemplated in this part of the ritual when you are summoning a being that is from an unclear sphere of origin. Again, we'll deal with this process later.

After the above preliminary link is established, it is time to give the conjuration to the spirit. This can be read from your book out loud, unlike the conjuration given in astral evocations, which must be memorized. After this is read, the book should be kept close by, as it will be used again to verify the identity of the spirit.

Once the conjuration is given, it is time to astrally travel to the entity's sphere of origin. This is done in the normal fashion explained earlier, so I won't repeat the steps of the process here. I will bring up the following difference, however: Before contemplating the disc or Tattwa to facilitate astral travel, spend a few more moments gazing at the sigil of the entity and repeating its name to yourself. Then begin your journey by gazing at the disc or Tattwa (see the "Ritual for Evocation to the

Physical Plane" below). Upon arriving in the sphere, initial contact has to be made with the entity. Simply calling out mentally to the entity should now bring it before you, as you have already conjured it.

When you behold the entity on the mental plane, you can then bring it back with you to the physical one. To do this, simply repeat its name as you journey back to the physical plane through the portal you entered from. Remember, the spirit has already been conjured and only needs a little guidance to pass through the planes and to manifest in the Triangle. The repetition of its name facilitates this, as it acts as a kind of homing signal for the spirit to follow. When you feel you have returned to your physical body, start repeating the spirit's name out loud, first as a whisper and then increasing in volume as you slowly open your eyes. The entity will then be standing in the Triangle before you; however, it may take a few moments for its created "body" to become clear.

When you finally have an entity standing before you in the Triangle, it is time to make sure you have evoked the right one. In evocations to the physical plane, the best way to do this is to ask it its name. If it answers correctly, then you should use your Sword to slide your open book across the floor into the Triangle, making sure you do not let your arm pass the confines of your magic circle. (In evocations to the physical plane, the Triangle is placed on the floor.) Then ask the entity to sign its name in your book with an astral impression (as explained in chapter 6). If the being signs its name correctly into your book, then you can proceed safely. If it doesn't sign its name, then you have the option to question the entity about its motive in coming to you or dismiss the entity and try again.

Once you are sure of the spirit's identity, give it a formal greeting like the one given in the evocation ritual below. After this, you may speak to the entity in the same way you spoke to entities summoned to the astral plane. Remember to keep your communication limited to topics relating to the purpose of your evocation, as the rambling of spirits can be confusing at times. The spirit may give you an easier way to evoke it in the future during this part of the ritual; if it does, make sure to write it down.

After you explain your reason for summoning the entity and are convinced the entity will be able to carry out your wishes, it is time to

Planet/Sephirah	Symbol	Disc Color	Symbol Color
1. Uranus/Kether	♅	White	Black
2. Neptune/Chokmah	♆	Gray	White
3. Saturn/Binah	♄	Black	White
4. Jupiter/Chesed	♃	Blue	Orange
5. Mars/Geburah	♂	Scarlet red	Green
6. Sun/Tiphareth	☉	Yellow gold	Violet
7. Venus/Netzach	♀	Emerald	Red
8. Mercury/Hod	☿	Orange	Blue
9. Moon/Yesod	☽	Violet	Yellow

TABLE 8.1

Sphere	Color	Incense	Divine Name	Metal	Wood	Gem	Creature
1. Uranus/Kether	White	Ambergis	Eheieh	Electrum	Aspen	Diamond	God
2. Neptune/Chokmah	Gray	Musk	Yah	Aluminum	Bramble	Turquoise	Man
3. Saturn/Binah	Black	Patchouli	Yhvh Elohim	Lead	Yew	Black onyx	Woman
4. Jupiter/Chesed	Blue	Cedar	El	Tin	Cedar	Amethyst	Unicorn
5. Mars/Geburah	Scarlet red	Peppermint	Elohim Gibor	Iron	Ash	Garnet	Scorpion
6. Sun/Tiphareth	Yellow gold	Frankincense	Yhvh Eloah Vedaath	Gold	Pine	Citrine	Lion
7. Venus/Netzach	Emerald	Rose	Elohim Tzabaoth	Copper	Elder	Emerald	Lynx
8. Mercury/Hod	Orange	Gum mastic	Shaddai El Chai	Quicksilver	Birch	Carnelian	Jackal
9. Moon/Yesod	Violet	Jasmine	Adonai Ha Aretz	Silver	Willow	Quartz	Elephant
10. Fire	Red	Olibanum	(see figure 2.1)	Iron	Oak	Fire Opal	Lion
11. Water	Blue	Myrrh	(see figure 2.1)	Silver	Willow	Aquamarine	Eagle
12. Air	Yellow	Galbanum	(see figure 2.1)	Aluminum	Hazel	Topaz	Bird
13. Earth	Black	Storax	(see figure 2.1)	Lead	Cypress	Salt	Bull

TABLE 8.2

give the entity a formal, carefully worded command, as you did during astral evocations. Once the entity carries out your wishes, if you have asked it to give you some specific information or carry out some task for you, you can dismiss it.

Dismissing an entity from the physical plane is similar to banishing an entity from the astral plane. A verbal dismissal is given, which clearly instructs the entity to harm no one as it leaves and to come quickly when called again, and the entity simply fades away. The performance of the banishing rituals in the Closing by Watchtower ensures the complete banishment of the entity from the room.

Preparing the Temple Room

The temple room used for an evocation to the physical plane has to be prepared in a special way before each ritual. As I mentioned earlier in the book, it is very difficult for an entity to manifest physically because of its etheric nature, which makes it alien to our own physical plane. To help an entity manifest, the temple room has to be made to vibrate in a similar way to the spirit's own sphere. When I say the room has to vibrate, I mean it has to give off influences which correspond to the entity's nature.

Preparing a room in this fashion can be accomplished in many ways. First of all, the color of the light used in the room has to be the same as the color of the entity's sphere. If you made the filter-stand for your lamp described in chapter 4, then this will not be a problem. Simply attach a sheet of cellophane to the filter that corresponds to the color of the entity's sphere. To find out which color to use, see table 8.2. Here you will find appropriate correspondences of all kinds to make your temple room agree with the nature of elemental and planetary workings. Other steps you can take to make your temple agree with the nature of the spirit you are trying to evoke are given below.

Several items should be placed on the Triangle in which the spirit will manifest. In evocations to the physical plane, this Triangle is placed on the floor, about two feet away from the circle's perimeter in the quarter that the evocation will be performed in, with the apex pointing away from the circle. The most important object that should be placed in the Triangle is the sigil of the entity. This will help draw the entity to the Triangle and will provide a focus for the entity's manifested energies. Therefore, in physical evocations, sigils can be charged in a unique way (see below).

The next most important item, which must be placed within the Triangle, is the censer. Burn incense in the censer that corresponds to the nature of the entity (see table 8.2). The rising smoke will do more than just help draw the entity to the Triangle, it will give the spirit some substance to help form a body around, in addition to the numerous particles it will pull out of the air for this purpose. If it looks like the censer is running out of incense during the ritual, you can balance some incense on the blade of your Sword and add it to the censer without leaving the circle. Make sure there is a good amount of smoke rising from the censer at all times during the evocation.

The rest of the items in the correspondence tables are optional and do not have to be added to the Triangle of Art. However, when you are trying an evocation to the physical plane for the first time, you may want to place as many of them in your Triangle as possible, as their vibrations will aid you in your working. In the correspondence table you will find gems, creatures, metals, and other objects that correspond to each sphere. Feel free to add as many of these objects, or representations of them, as you like to the Triangle. When I say "representations" of objects, of course I mean those things listed in the tables that you cannot possibly place in the Triangle. For example, you should use a statue of a lion, as a real one may become rather difficult to control during a ritual!

Preparing Sigils

As I hinted earlier, sigils can be charged in a special way in evocations to the physical plane. In these types of evocations, two sigils are used: one that the magician gazes at, and one that is placed within the Triangle. The latter, therefore, comes into physical contact with the entity evoked. Because of this, the sigil within the Triangle is automatically charged with some of the energy the spirit expends toward satisfying your wishes. Keeping this sigil will magnify the effects of your evocation. For example, if you were to summon a spirit to help you win big at a casino, carrying the charged sigil of that spirit in your pocket would focus that magical "good luck." These sigils are more powerful than those used in astral evocations, which only help establish links with entities, making them less effective due to their lack of physical contact with the entity.

Once the goal of an evocation is accomplished, the sigil can be destroyed to release its energy. Again, burning the sigil is the best way to facilitate this, making paper the ideal material to use when constructing sigils you want to be able to get rid of easily. Of course, you would only use paper if you wanted to destroy the sigil in the future. Sometimes, however, you may want to assign a spirit to a task which it may take months, years, or perhaps, even lifetimes to complete. Tasks like this are not unheard of; actually, they were quite common in the Middle Ages, when magicians would command spirits to guide armies to victory in decade-spanning wars or provide plentiful harvests for their people. The effects of this type of magic would be felt for years after a magician's death and, quite possibly, there are still spirits carrying out the wishes of magicians who lived hundreds of years ago!

What kind of long-range tasks would you want to assign to a spirit? The most common type of long-range task to assign a spirit is protection. By commanding a spirit to protect a certain place (i.e., your magical library), and then placing its charged sigil on the premises, you are ensuring that area will be protected indefinitely from whatever type of harm you specify. If you ever wish to assign a task like this to a spirit, then you may want a sigil that is a little more durable than paper.

In table 8.2, you'll notice that one of the columns is a listing of appropriate metals for each of the spheres. If you have access to an engraving gun (you can find one for as little as twelve dollars), then you may want to try engraving the sigil of a specific entity onto its corresponding metal. You can still make the one you gaze at out of paper, of course, as the metal one is the one that goes into the Triangle, and consequently, the one that becomes charged. Finding some of the metals might be difficult, and if this is the case, you can carve the sigil into a corresponding type of wood (again, see table 8.2) and apply a coat of clear finish to protect it. Either way, you'll end up with a sigil that should last for years.

Once you have a more or less permanent sigil, and you charge it by performing an evocation, you can then place it where its influence will have the greatest effect. With a little bit of creativity, you can even turn a sigil into a necklace. This is especially helpful if you command a spirit to protect you from physical harm. The sigil necklace would then act as an amulet to ward off this type of evil. Necklaces could be made to protect from disease, help in business matters, and even improve your foreign language skills. I'm sure you can see the countless possibilities.

Ritual for Evocation to the Physical Plane

Now, let's move on to the actual ritual for evoking spirits to the physical plane. It is given below in a form that makes it suitable for calling spirits from either elemental or planetary spheres. To evoke beings from unclear spheres of origin, consult the modifications to the ritual and the discourse that follows the ritual below.

The Divine Names used in the conjuration below are not given. Instead, you should use the name of God that corresponds to the sphere you are working with (see table 8.2). The places where Divine Names should be inserted are indicated with a "D.N." to avoid confusion. In the case of beings from unknown spheres, feel free to use whatever Divine Names you feel comfortable with. In addition to the insertion of appropriate Divine Names into the ritual, there are also places where you will have to add the sphere's name (indicated by "S") and the entity's title (if the entity has a title, replace the word "Spirit" with it). All of these factors will help personalize each conjuration, and consequently increase its potency.

Inserting the names of the elemental regions should pose no problem; however, the interchangeable nature of the names of the Sephiroth and planets could create some confusion. To prevent this from happening, use the correct planet name for entities that are obviously planetary in nature and the name of the correct Sephirah for beings such as angels, which obviously come from the Sephiroth. For example: Och, Ruler of Sol (the Sun); or Archangel Raphael of Tiphareth.

1. Prepare your temple room as usual with the following additions. Position your Triangle on the floor in the correct quarter of the room, two feet outside your circle, with its apex pointing away. Place within it the sigil of the spirit, the burning censer, and any other corresponding items you have. Your lamp should be placed on the floor, covered with a filter of an appropriate color, and adjusted to fill the room with colored light. Have your book, second sigil (which should be wrapped in black silk until needed), and either disc or Tattwa present on your altar. Make sure you have extra incense to add to the censer with your Sword if needed (see above). Also, wear your robe and ring if you have them.

2. Perform the Opening by Watchtower up to and including step 17 (see chapter 3).

3. Unwrap your sigil and place it and your disc or Tattwa on top of your book. Hold this little bundle in your left hand and pick up your Magic Wand or Sword (depending on the nature of the working) with your right. If your Sword is not being used, have it ready in case you want to add incense. Move clockwise to the quarter you will be calling the spirit from.

4. Spend a few moments gazing at the Tattwa or disc and then at the sigil of the entity. Repeat the entity's name to yourself a few times.

5. Open your book to your conjuration and read it, or the one following, out loud, making sure to correctly substitute the necessary names: *I evoke and conjure thee, O Spirit N.,* (replace "Spirit" with the entity's title, if any), *by the power and authority of the Supreme Majesty—the true God Who is known by the name of D.N.* (substitute and vibrate the Divine Name of the sphere) *in the realm of S.* (name of sphere)—*to come quickly from that realm and to*

*appear before me within this Triangle of Art, in a fair and
comely shape.*

6. Rest your book on the altar behind you, but keep it
    within reach. Hold your sigil over your Tattwa or disc
    and intently gaze at the seal of the entity in your hand.
    Once again, repeat the name of the entity you are calling
    a few times.

7. After a few moments, place the sigil of the entity under
    the Tattwa or disc and gaze at the latter. If you are using
    a Tattwa, allow the complementary image of it to fill
    your mind's eye. In the case of the planetary disc, try to
    recall its color and symbol.

8. Still standing, with your sigil and disc or Tattwa in your
    left hand and your Wand or Sword in your right, astrally
    travel to the entity's sphere of origin by passing in your
    astral body of light through a portal you create.

9. When present in the correct sphere on the mental plane,
    vibrate the appropriate Divine Name. After doing this,
    call out to the entity you have already conjured.

10. Once the entity appears to you, begin mentally repeating
    its name as you return to your physical body. When you
    arrive, keep your eyes closed and continue to mentally
    repeat the name of the spirit.

11. When you feel fully in control of your physical body once
    more, begin whispering the name of the entity out loud.
    With each repetition, let the volume of your voice increase,
    and slowly begin to open your eyes. You should see the
    entity manifesting in the smoke of the censer.

12. Once the entity has fully manifested, ask it what its
    name is. If it followed you to the physical plane using

your beacon, then it will most likely be the correct spirit. It pays to be certain, however. If the spirit answers correctly, move on to step 13. If it claims to be some other spirit, then you can either question it or banish it.

13. Open your book and slide it across the floor into the Triangle using your Sword. Ask the entity to sign its name as an astral impression in your book.

14. Retrieve the book with your Sword. If the name is legible and signed correctly, give the entity a greeting you wrote yourself or the following one: *In peace I welcome you, Spirit N.* (again, substitute title, if any), *and in the name of the Most High I command you to remain within this Triangle of Art until you are dismissed, to speak honestly and to truthfully answer all questions put before you.*

15. You may now freely converse with the spirit. Remember to stick to topics concerning the purpose of your evocation. If the entity gives you a ritual or word of power to call it with in the future, write it down in your book.

16. When you have established the entity's ability to carry out your wishes, give it a carefully worded formal command of your own creation, as you did in astral evocations.

17. After the entity answers whatever questions you put to it or agrees to carry out your wishes, you can dismiss it using a formal dismissal like the following: *Go in peace and return to S.* (name of sphere), *your sphere of origin, O Spirit N. By the authority of the True God D.N.* (vibrate the correct Divine Name), *I command you to harm none as you depart, and to be ready to come quickly if called again.*

18. After the entity vanishes, move clockwise to your position to the west of the altar, facing east. Close your book and recover the sigil with black silk.

19.   Perform step 18 of the Watchtower Ritual.

20.   Perform the Closing by Watchtower.

After completing an evocation, always perform some type of grounding activity, such as taking a walk or eating something. If there is an important piece of information to record, do so before you ground yourself. Do not, however, write your general impressions of the evocation until after you have fully returned to a non-magical state of consciousness.

Calling Entities with Unclear Spheres of Origin

Once you have succeeded at physically evoking entities with known spheres of origin, you may want to try calling an entity with an unclear one. There are quite a few of both of these types of spirits listed in chapter 9. What follows is a method for evoking these seemingly "wandering" spirits to the physical plane.

Note: You do not need to know a spirit's plane of origin to evoke it to the astral plane, so any spirit could be called in that fashion using the ritual in chapter 7. The technique given below is only for use in physical evocations.

Spirits that do not come from specific spheres of origin are very common in grimoires. The *Goetia,* for example, contains nothing but these types of spirits. Whenever you find one of these beings that you wish to evoke to the physical plane, you are faced with the dilemma of trying to contact a spirit that could be anywhere in the magical universe. How does one go about accomplishing such a task?

Actually, the process for evoking these spirits is not at all difficult, just different. If you have already succeeded at other physical evocations and are, as a result, adept at astral traveling, then you should have no problem using this technique. It is nothing more than a variation on what you have already done.

First of all, you must be able to travel to an area of the mental plane where the spirit currently dwells. You will need the spirit's sigil to accomplish this. Gaze at the sigil as you have done before, and repeat the entity's name. Do this before and after you give the conjuration to the spirit.

The second time you contemplate the sigil, try to memorize its shape. Then, when you close your eyes, project this shape onto an imaginary curtain in the air above you. Once you do this, enter your body of light and travel through the curtain. On the other side of the curtain, you will find yourself in close proximity to the entity. Bring it back with you by repeating its name and continue the evocation in the normal fashion.

With a little practice, the above technique should work very well for you. Once you master it, you can take pride in knowing there isn't a single spirit in the magical universe you can't evoke to the physical plane. I need not reiterate the powerful possibilities available to you once you achieve this level of magical development.

Chapter 9 contains the descriptions and sigils of several entities. Remember, these are all beings I have personally evoked and worked with, so you can be certain they will accomplish whatever tasks I have listed. These spirit listings are included to help get you started quickly in either form of evocation. Of course, should you decide to acquire one or more grimoires, you will be able to evoke those entities just as easily, although you should be wary of some of the promises made in these books, as I have found from experience that they are not always accurate.

In addition to evoking pre-existing spirits, you may want to try the advanced technique given in chapter 10. This chapter will teach you how to create and work with your own egregore. Having come this far, you will have no trouble performing this magical technique.

chapter nine

FIFTY ENTITIES
TO INVOKE

Djin, the King of Fire, can explain several
uses of his element in magical workings.
Particularly, he can teach you how to empower any rituals of a Fiery nature with
the elemental essence needed for success.

Below you will find the names, descriptions, and sigils of various types of entities. These spirits are grouped according to
their spherical origins to make it easier to
locate the type of being you would like to contact. Three categories are used to classify the
spirits in this manner: Spirits of Elemental
Spheres, Spirits of Planetary or Sephirothic
Spheres, and Spirits of Various Regions. Each
of these categories has already been explained.

As a whole, this listing contains fifty spirits of several different types taken from various sources. Many of the spirits listed here will
be familiar to those who have read some of the

more famous grimoires, while some of the entities described below appear here in print for the first time, as they are entities I have encountered while traveling on their respective planes.

I have found that on occasion, entities appear in slightly different forms, some of which are not very pleasing to behold. If a spirit ever appears in a form that disturbs you, simply command it once more to take on a more pleasing form and, as long as it is not a demonic entity, it should generally obey this command. Some benevolent spirits cannot change their forms, however, and it will be up to you to decide if you want to work with them or not. The spirits given below all have appearances that I have been able to handle, but if you feel extremely uncomfortable with a particular spirit's appearance, and it is not able to change its form, then you should consider banishing the entity and evoking a different one.

The sigils of the entities are found in groups on pages close to the descriptions of the entities they represent. Wherever possible, I have given the traditional sigil of each entity. When these were not available, I have provided a derived sigil, using the Hebrew Rose Cross Lamen for angels, and the Roman version of the Lamen for entities with non-Hebrew names.

To make certain the entity listings given below are easily understood, following is an explanation of each part.

Number/Name: The name of the entity is given, along with a number to make it easy to find at a later time.

Title/Type: If the spirit has a title (i.e., King), it is given here. Also, if the spirit is of a specific type (i.e., angel), it is listed here. If the spirit is of no particular type and has no title, then NA (not applicable) will be given.

Sphere: The specific sphere of origin for the entity is given. In the case of Elemental Spheres, either Fire, Water, Air, or Earth. In the case of Planetary or Sephirothic Spheres, entities of the two types are distinguished to make it easier to modify conjurations (see chapter 8). For example, either Venus or Netzach will be given for entities of this type. If the entity does not come from a specific sphere, then NA is given.

Appearance: A description of the entity's appearance is given here.

Expertise/Tasks: Here you will find the areas a spirit is knowledgeable in, as well as various types of tasks it can perform. I have only listed subject areas or tasks I have personally found the entity to be proficient with. Some of these beings may have talents I have not yet discovered, and for this reason you are encouraged to experiment or interview these beings to find out any other functions they may serve. Whether you choose to experiment or not, however, you can at least work with the assurance that whatever abilities are listed in this section of an entity's description are accurate.

Spirits of Elemental Spheres

Fire

1. Djin

Title/Type: Elemental King

Sphere: Fire

Appearance: This spirit looks somewhat elderly, yet has great muscles. He is dressed in a red robe that drapes over his right shoulder. Djin has flames for hair, which constantly move around his heavy-looking, golden crown.

Expertise/Tasks: This King of Fire can explain several uses of this element in magical workings. Particularly, he can teach you how to empower any rituals of a Fiery nature with the elemental essence needed for success. If you evoke him, you will also learn many new types of Fiery magical rituals you didn't even know existed, all of which he will make you promise never to reveal to another. Like other elemental kings, Djin can give you the names and sigils of several Elementaries in his realm that you may want to summon.

2. Seraph

Title/Type: Elemental Ruler/Angel

Sphere: Fire

Appearance: Seraph usually has a fierce expression on his smooth, unlined face. He is clothed only in golden-yellow flames, and has a lean but well-developed physique. It is almost impossible to avoid looking at his blazing eyes, which seem to be criticizing your every move.

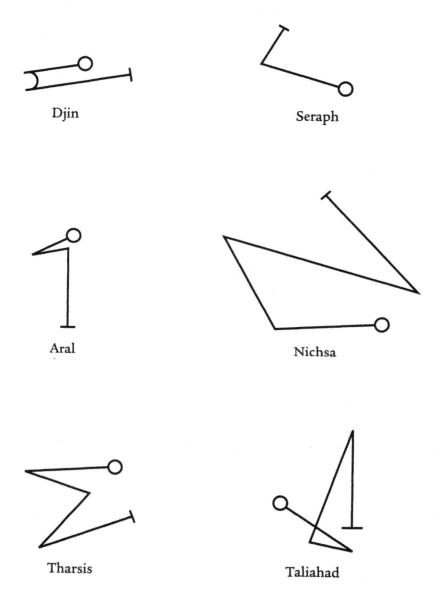

Djin

Seraph

Aral

Nichsa

Tharsis

Taliahad

SIGIL SHEET 1

Expertise/Tasks: One of the most potent things this entity can teach is the manipulation of Elemental Fire within one's body. As a result, you can give yourself energy to accomplish whatever tasks you wish, whether they are magical or not. He can also teach you how to make several potions to help you control this energy within yourself or others. Like Djin, Seraph can tell you of many salamanders you can summon to accomplish a variety of purposes.

3. Aral

Title/Type: Angel

Sphere: Fire

Appearance: Aral has a gaunt face and wears a light-blue robe. On his head he wears a light-blue winged helmet. He has a masculine body, though not as muscular as the above Fire beings. Aral has large golden wings, which when extended are amazing to behold as you can see the shadows of flames dancing on them.

Expertise/Tasks: This angel has the ability to control the physical manifestation of Elemental Fire and can teach you how to actually manipulate heat and flames through the use of special rituals. There are a few requirements you must meet before you can apply this technique, however, and Aral will explain this to you when you evoke him.

Water

1. Nichsa

Title/Type: Elemental King

Sphere: Water

Appearance: Nichsa, like all Water beings, is pleasing to behold. This king, in particular, appears youthful and has blonde hair and blue eyes. His crown looks as if it is made of glass and his robe looks like it is made of a constantly swirling blue liquid. Staring at the robe is quite disorienting.

Expertise/Tasks: This King of Water can teach you how to use Elemental Water in rituals to empower them. He can also instruct you in the amazing practice of controlling the weather, particularly rain and fog.

2. Tharsis

Title/Type: Elemental Ruler/Angel

Sphere: Water

Appearance: Tharsis seems to be almost androgynous. The face and upper body seem to be masculine, yet it is unmistakably feminine from the waist down. The robe of this being is made of a yellow transparent material.

Expertise/Tasks: Tharsis is extremely helpful in matters of love and relationships. This ruler can teach you how to work Watery magic, which ensures peace in all kinds of relationships, and can explain the truth behind the concept of soulmates.

3. Taliahad

Title/Type: Angel

Sphere: Water

Appearance: Taliahad appears as an exceptionally beautiful young woman with completely purple eyes that are dangerously hypnotic to look into. Her skin is a smooth and flawless ivory color, and she is naked except for a gray sash tied around her waist. (*Note:* Reread the warning given in the introduction about not allowing yourself to become fascinated with beings of this type.)

Expertise/Tasks: Taliahad is extremely knowledgeable in the art of talismanic magic. She can teach you how to create and charge talismans to accomplish just about any purpose. If you desire help in finding a suitable mate for yourself or someone else, Taliahad can actually guide you through a magical talismanic ritual to accomplish this. If you ask her about this ritual, she will tell you the appropriate time to evoke her again to perform it effectively.

Air

1. Paralda

Title/Type: Elemental King

Sphere: Air

Appearance: This king appears as an armored knight surrounded by clouds. He has wings on his helmet and boots, and his armor is made of golden-yellow metal. Paralda's face gives one the impression of deep thought.

Expertise/Tasks: Paralda is an excellent teacher in the area of learning new concepts. He can show you how to absorb new ideas and teach you how to create an elixir that will help strengthen your control over the Airy mental faculties. Paralda's control of the mental processes also includes that of telepathy, and he can teach you how to master this psychic ability.

2. Ariel

Title/Type: Elemental Ruler/Angel

Sphere: Air

Appearance: Ariel appears as a thin, almost feminine-looking man. He has golden-colored wings and wears a gray robe with a golden-yellow belt. A bright aura surrounds his head.

Expertise/Tasks: Ariel can teach several spiritual and natural sciences. He is a storehouse of theoretical as well as practical knowledge. One of his most amazing abilities is to teach the magician how to use almost any form of divination successfully.

3. Chassan

Title/Type: Angel

Sphere: Air

Appearance: Chassan appears androgynous, with a wide, feminine face, and thin, male-looking body that is clothed with a long, brown skirt.

Paralda

Ariel

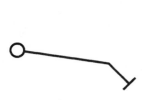

Chassan

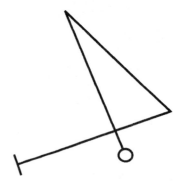

Ghob

Kerub

Phorlakh

SIGIL SHEET 2

Expertise/Tasks: Chassan can teach you how to control the physical manifestations of Elemental Air, particularly wind. By working with Chassan, you will learn rituals that will help you manipulate winds to amazing degrees. Also, rigorously practicing the rituals you learn may enable you to master levitation by changing the air pressure above and below you, as Chassan will explain.

Earth

1. Ghob

Title/Type: Elemental King

Sphere: Earth

Appearance: Ghob appears as a short, bearded man with dark hair and luminescent eyes that glow a dark green color. He is dressed in brown robes and wears a necklace with a large, black crystal pendant.

Expertise/Tasks: This King of Earth can teach you how to control the forces of this element in magical rituals. He can also show you how to call forth various gnomes to help you in underground endeavors such as treasure hunting and archaeological excavation. Aside from teaching these important magical techniques, Ghob can also explain what true alchemy is and teach you how to apply the science to your life.

2. Kerub

Title/Type: Elemental Ruler/Angel

Sphere: Earth

Appearance: This ruler is somewhat short and heavy, yet very muscular. He has unusually thin legs and wears a violet robe. Kerub's face appears to be chiseled from stone, as his serious expression almost never vanishes. He wears several rings with large gems on his fingers.

Expertise/Tasks: Kerub is an excellent teacher of the mystical properties of gems. He can show you ways to use crystals and other types of gems to accomplish almost any magical goals you can think of. This ruler is also knowledgeable in the divinatory art of geomancy and can teach you how to use it successfully.

3. Phorlakh

Title/Type: Angel

Sphere: Earth

Appearance: Phorlakh is a fairly tall, androgynous angel. It has the face and genitals of a woman, but a strong muscular body of a man. It is clothed with only a dark purple sash over its right shoulder.

Expertise/Tasks: Phorlakh is skilled in the practical magical use of Elemental Earth energies. If you desire help in acquiring money, or other forms of abundance or stability, then this angel can help you by showing you how to perform simple rituals toward these ends. You can also assign Phorlakh to the task of finding out where you should look for jobs or opportunities.

Spirits of Planetary or Sephirothic Spheres

As I mentioned earlier, the spherical nature of each entity given will be clarified in each case. Note that the heading for Malkuth does not include a corresponding planet or planetary entities. This is because the planetary correspondence of this sphere consists of the Four Elements, the inhabitants of which were already explained. Therefore, only the Sephirothic Archangel for Malkuth is given.

Uranus/Kether

1. Metatron

Title/Type: Archangel

Sphere: Kether

Appearance: An incredible figure to behold, Metatron appears as an enormous being of brilliant white light. Metatron's

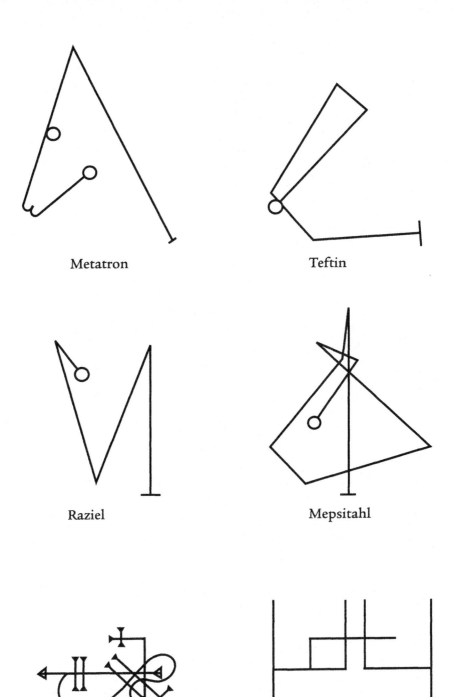

Metatron

Teftin

Raziel

Mepsitahl

Tzaphqiel

Aratron

SIGIL SHEET 3

features are hard to distinguish due to his luminosity, but you can clearly make out his outline from the chest area up. He holds a scroll in his hand.

Expertise/Tasks: This archangel should only be summoned if you wish to follow a path leading to true adepthood. Metatron can teach you how to work toward the completion of the Great Work and can help you find a suitable teacher or magical group to work with.

2. Teftin

Title/Type: NA

Sphere: Uranus

Appearance: Teftin appears as an elderly and wise-looking man with long white hair. In his left hand he carries a book and in his right he holds a large black tablet that flashes images. His robe seems to be made of a metallic cloth.

Expertise/Tasks: If you are interested in technology, this is the spirit to work with. Teftin is able to help you develop new technologies and to repair or improve existing ones. Gazing into his tablet will provide you with glimpses of several future devices, and even if you are not at all interested in developing any of them, learning about our technological future is still extremely fascinating. Writers of science fiction may want to summon this spirit to help them come up with new and realistic ideas.

Neptune/Chokmah

1. Raziel

Title/Type: Archangel

Sphere: Chokmah

Appearance: Raziel's head is surrounded by a glowing yellow aura, which makes it difficult to make out his face. He wears a robe of gray material that seems to swirl as if made of liquid. Raziel appears quite tall and has large sky-blue wings. In his hands he holds a large gray book.

Expertise/Tasks: This archangel is knowledgeable in several mystical arts and can explain the truth behind many occult truths in the universe. One of the most interesting things Raziel can give you is an understanding of the energy currents in the magical universe and how to manipulate them to various ends.

2. Mepsitahl

Title/Type: NA

Sphere: Neptune

Appearance: This spirit appears as a young woman with white hair and green eyes. She wears a blue-green robe and a headband of the same color with a purple gem in its center that covers her third-eye area.

Expertise/Tasks: Mepsitahl can teach you how to master several occult faculties such as telekinesis, clairvoyance, and telepathy. The exercises she will teach you are unlike any you have ever practiced before, and if practiced regularly will yield excellent results. Make sure you have your book handy to take notes of what Mepsitahl tells you, as she has a wealth of information to share.

Saturn/Binah

1. Tzaphqiel

Title/Type: Archangel

Sphere: Binah

Appearance: This angel is somewhat intimidating to behold. He appears as a tall man wearing a black robe. On his head he wears a blue Egyptian nemyss, and around his neck, a red amulet. He often carries a dark rod in his right hand and a glowing cup in his left. His wings are bright silver in color.

Expertise/Tasks: This archangel is omniscient in that he has seen everything that has ever occurred. As a result, he can tell you of many hidden truths, but must first be convinced of your need to know them. Make sure your intentions in

asking Tzaphqiel's help are pure, because he will be able to
see through you if you are lying.

2. Aratron

Title/Type: Olympic Planetary Spirit

Sphere: Saturn

Appearance: Aratron appears as a slender, bearded man riding
on a black dragon (if this disturbs you, ask him to dis-
mount and he will). His robe is made of a black material
that is covered with multicolored gems which seem to glow.

Expertise/Tasks: This Olympic Spirit has the ability to transmute
elements. He can petrify organic tissue and change stones
into different gems. If you summon him to the physical
plane, he will do the latter for you, but the durations of his
manifestations vary, as he will explain. In addition to causing
external changes, Aratron can help you make changes within
your subtle bodies that improve your health, allow you to
perform magical rituals with greater success, and help you
complete the Great Work of reuniting your consciousness
with the Divine. The latter is the true goal of alchemy. Aratron can also give you the names and sigils of several entities
that can help you perform alchemical magic.

3. Harayel

Title/Type: Angel

Sphere: Saturn

Appearance: This angel appears as a young, red-haired woman
dressed in a dark brown robe. She has golden wings and a
blue belt in which a short sword rests.

Expertise/Tasks: Harayel can teach you quick and effective rituals for providing protection for yourself in any of the three
planes. If you ever feel you are in psychic or physical danger,
practicing one of these rituals will immediately bring you
peace.

Jupiter/Chesed

1. Tzadqiel

Title/Type: Archangel

Sphere: Chesed

Appearance: Tzadqiel appears as a white-winged figure, wearing a blue robe. He has golden hair and blue eyes.

Expertise/Tasks: Tzadqiel is best described as a wise ruler. He can help you make decisions of varying importance and help you diplomatically handle situations in your life. When confused or distressed about a difficult decision, you will find this archangel's advice brings great peace.

2. Bethor

Title/Type: Olympic Planetary Spirit

Sphere: Jupiter

Appearance: Bethor appears as a man wearing a deep azure robe and a bishop's miter on his head. Flashes of light seem to dance around this spirit, as if a lightning storm was in the distance behind him.

Expertise/Tasks: Bethor is a very helpful spirit to work with as he can help you get a more prestigious job and can tell you the location of hidden treasures. The spirits he can assign to work for you can actually guide you in your everyday life. Once you contact Bethor, pay close attention to your surroundings when you are dealing with business or economic concerns, and you will often become aware of the proximity of one of his spirit servants.

3. Riprinay

Title/Type: NA

Sphere: Jupiter

Appearance: When you first summon Riprinay, he will appear as a large blue peacock. If you ask him to change his appearance, he will turn into a blue cloud and reform as a man with the head of a peacock, wearing a golden robe.

Harayel

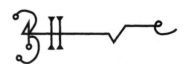

Tzadqiel

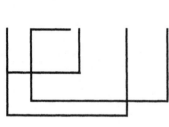

Bethor

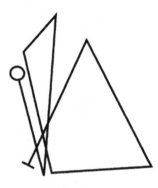

Riprinay

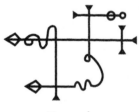

Kamael

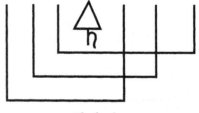

Phalegh

SIGIL SHEET 4

Expertise/Tasks: Riprinay has the amazing ability of seeing into the future to predict economic changes. He can help you plan which stocks to invest in and can foresee market conditions that will give you an investment edge in other areas as well.

Mars/Geburah

1. Kamael

Title/Type: Archangel

Sphere: Geburah

Appearance: Kamael appears as a warrior dressed in a red tunic. He wears green armor plates and an iron helmet. He carries a scale of justice and has large green wings.

Expertise/Tasks: The most impressive trait of this archangel is his knowledge of karma. Kamael can show you how to make up for bad karma and, consequently, how to purify yourself in this incarnation. This archangel also acts as a kind of judge and can act on your part to make sure justice is served. Kamael's aid can be particularly helpful if you are involved in some type of legal dispute, but he will only help if he feels you are honest.

2. Phalegh

Title/Type: Olympic Planetary Spirit

Sphere: Mars

Appearance: Phalegh appears as a strong warrior with blazing eyes. He wears a red tunic and carries a long iron sword.

Expertise/Tasks: If you need courage, strength, or energy to perform some difficult task, then call upon Phalegh and he will bestow these traits upon you. If your profession requires you to exhibit courage and internal strength (i.e., soldier or police officer), then this spirit can help you perform your duties without difficulty. But for those of us in more civilian roles, Phalegh's energies are still very helpful.

3. Sartmulu

Title/Type: NA

Sphere: Mars

Appearance: This spirit appears as a tall, muscular man, wearing an armored skirt. His long dark hair seems to be sculpted from stone.

Expertise/Tasks: Sartmulu can teach you how to exercise your body physically and magically to become strong and healthy. People who aspire to become weight lifters or bodybuilders will be very pleased with the results of working with this benevolent entity.

Sun (Sol)/Tiphareth

1. Raphael

Title/Type: Archangel

Sphere: Tiphareth

Appearance: In his aspect of Archangel of Tiphareth, Raphael appears in a blaze of golden light. His body seems to be a three-dimensional extension of the light itself.

Expertise/Tasks: Raphael is often referred to as the Divine Physician. If you are ill, or know of anyone who is, calling on Raphael may help that person get better. Keep in mind that for karmic reasons, sometimes healing magic cannot help a person who needs to experience a certain disease. This archangel will explain all of this to you when you evoke him. Raphael is also very knowledgeable in Hermetic Science and can act as a teacher in this discipline.

2. Och

Title/Type: Olympic Planetary Spirit

Sphere: Sol

Appearance: Och appears as a crowned man, wearing golden robes and riding on a lion. In his right hand he holds a scepter, which gives off brilliant golden light that is almost blinding to view.

Expertise/Tasks: The wisdom of this spirit is fascinating. He can tell you of many secrets in the world and show you how to heal others. Och also has an amazing ability at working with gold and can show you how to work with this substance in an occult manner unknown to most individuals.

3. Menqel

Title/Type: Angel

Sphere: Sun

Appearance: Menqel appears as a woman with green eyes. She has a long, dark purple robe and gray wings.

Expertise/Tasks: Menqel is a great peacemaker whose power influences people and even animals. In addition to helping you establish peace in your life, Menqel can also help you find harmony in your career. If your workplace is an uncomfortable place to be, you might want to wear a charged sigil of this angel to help you get through the day.

Venus/Netzach

1. Haniel

Title/Type: Archangel

Sphere: Netzach

Appearance: Haniel appears as an androgynous figure with large gray wings. This angel is dressed in an emerald green robe and carries a brown lantern.

Expertise/Tasks: Haniel is the archangel of love and harmony and can help you bring these influences into your life. In addition to being helpful in matters of the heart, Haniel can also aid you in artistic matters. If you need harmony and inspiration in your life, this is the archangel to work with.

2. Hagith

Title/Type: Olympic Planetary Spirit

Sphere: Venus

Appearance: This spirit appears as a beautiful woman riding a camel. Hagith is naked except for a green sash tied around her waist. She often holds flowers in her left hand.

Expertise/Tasks: Hagith is very helpful in matters concerning love and beauty, as are many Venus entities. When working with Hagith, you are able to also call upon her servant spirits, which will guide you in matters of the heart in your everyday life. Under no circumstances, however, will Hagith or her servants perform any type of mind-controlling magic. Instead, Hagith will effect changes within you that draw love into your life. She also has the amazing ability of making infertile men and women fertile again. In addition, Hagith can control the fertility of the earth, which is very helpful to farmers.

3. Amsariah

Title/Type: NA

Sphere: Venus

Appearance: This spirit appears as a young man wearing a green robe that is Roman in appearance, as it hangs over his shoulder. He has short blond hair and green eyes and carries a harp made of horns.

Expertise/Tasks: Amsariah is an excellent spirit to summon if you are a musician or any other type of artist. This spirit can help you create masterpieces of art in any medium you wish. He also knows how to make an herbal potion that stimulates creativity. When asking him for this formula, however, be persistent as Amsariah tests the determination of the magician who summons him. The first formula he gives you for the potion may contain some herbs you have never heard of. If this occurs, make him aware of your discovery of his intentional error, and he will give you the correct formula.

Mercury/Hod

1. Michael

Title/Type: Archangel

Sphere: Hod

Appearance: As Archangel of Tiphareth, Michael appears as an angel with wings, in orange and emerald robes. In his right hand is a long spear.

Expertise/Tasks: Michael is one of the most powerful entities of protection in the universe. Whether you need advice on this subject or actual physical protection, Michael is more than eager to aid you. This Archangel is also very knowledgeable in the art of evocation, and because of his other assignment as Archangel of Fire, his name is inscribed in the Triangle of Art. He can answer any questions you may have about advanced evocation techniques and even teach you new methods for evoking entities.

2. Ophiel

Title/Type: Olympic Planetary Spirit

Sphere: Mercury

Appearance: Ophiel appears as a young man wearing a robe that constantly changes color. In his right hand he holds a wand; seated at his left side is a white dog.

Expertise/Tasks: Ophiel is an excellent occult teacher. He can show you how to perform almost any type of magical ritual and help you establish a daily regimen of exercises that will help you advance as a magician. Also, Ophiel can discover the locations of occult books or implements you need but cannot find.

3. Hihaiah

Title/Type: Angel

Sphere: Mercury

Appearance: This angel appears as a middle-aged woman with red hair. She wears a gray robe and a red belt and carries a yellow scroll in her right hand.

Expertise/Tasks: There are many occult secrets in the universe that are hidden in plain sight through the use of symbolism. Hihaiah can show you how to find the true meaning of the occult symbols used in the world. Once you learn how to decipher these, you will be amazed to find how much practical occult knowledge is available, yet never explained in writing. If you request it, Hihaiah will also show you how to make potent magical symbols of your own to use in rituals. These will appear on her scroll when she unrolls it or, if you prefer, Hihaiah will draw them on paper you supply (of course, this is only possible if she is evoked to the physical plane). These astral impressions of symbols on paper will fade just as spirit signatures do, so you may want to trace them in ink during the ritual.

Moon (Luna)/Yesod

1. Gabriel

Title/Type: Archangel

Sphere: Yesod

Appearance: In his aspect of Archangel of Yesod, Gabriel appears as a winged angel with wavy silver hair in a purple and black robe.

Expertise/Tasks: Gabriel is a peacemaker whose influence can affect the entire planet. He is also an excellent teacher in the art of true skrying. By this I mean he can teach you how to view specific events in the past, present, or future, regardless of their location.

2. Phul

Title/Type: Olympic Planetary Spirit

Sphere: Luna

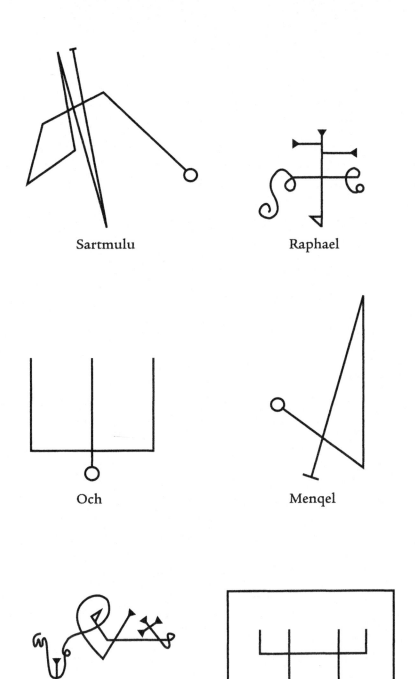

Sartmulu

Raphael

Och

Menqel

Haniel

Hagith

Appearance: Phul appears as a hunter, carrying a bow in her left hand. She wears a green tunic with silver armor and has a quiver of arrows on her back.

Expertise/Tasks: This spirit has the Lunar ability of controlling the waters of the earth. Phul can summon Undines that will perform any tasks you assign them (see chapter 1 for a listing of Undine abilities). In addition to controlling water, Phul can show you how to work with silver to create very powerful talismans. When she appears to you, be sure to also ask her for the method of true astral projection, as this practice will open several magical possibilities to you in the future, including a form of time travel.

3. Vimarone

Title/Type: NA

Sphere: Luna

Appearance: Vimarone appears as a man with long brown hair, wearing a purple robe. He carries a silver mirror in his right hand.

Expertise/Tasks: This spirit can teach you how to manifest objects on the physical plane by first creating them on the mental and astral planes. When you first practice this type of magic, you will find that it may take as long as a month to manifest the object you have concentrated on. With practice, however, you will find the time it takes to manifest an object will become shorter, and eventually, almost instantaneous.

Malkuth

1. Sandalphon

Title/Type: Archangel

Sphere: Malkuth

Appearance: Sandalphon appears as a tall angel, dressed in black. His robe seems to be made of pure energy and, if gazed at, it becomes evident that it is constantly emitting dark sparks of energy.

Expertise/Tasks: The primary function of this archangel is to act as a guide. If you need advice regarding communication with entities from Elemental or planetary spheres, Sandalphon can help steer you in the right direction. Also, if you are ever confused about the direction your spiritual path is taking you, contact Sandalphon for advice.

Spirits of Various Regions

1. Sirchade

Title/Type: NA

Sphere: NA

Appearance: This spirit is a little difficult to behold the first time you call him. Sirchade manifests as an amalgam of several different types of animals, with a canine head, bat-like wings, and a scaly body. Commanding this spirit to improve its appearance will result in it taking the form of a cloaked figure with a not so human face, as its mouth will still be somewhat elongated.

Expertise/Tasks: Sirchade has the ability of summoning and controlling animals of all types. He can teach you rituals and techniques for working similar feats and show you how to make an amulet to keep dangerous animals away from you. If you have a favorite pet, Sirchade will be able to show you how to establish a special link with that animal, making it a special type of familiar.

2. Bael

Title/Type: King/Goetic

Sphere: NA

Appearance: When I first summoned this spirit, he appeared as a large toad. I commanded him to take on a more human appearance and he did so. His face is rather feline when he is in human form and his ears are slightly pointed at their tops.

Expertise/Tasks: Bael can teach you how to become invisible. Like other magical techniques, the effects of invisibility are first only felt on the mental and astral planes. Invisibility on the mental plane causes people to ignore you if they are not looking for you. When mastered on the astral plane, however, it allows your presence to be undetectable, even if someone is actively searching for you. Invisibility on the physical plane, of course, will make you transparent. I've never managed to perform this last type of invisibility, which is to be expected, as Bael claims it takes years to master the art to this degree.

3. Frucisierre

Title/Type: NA

Sphere: NA

Appearance: This spirit appears as a tall, gaunt man with gray, wrinkled skin. He is dressed in a black Arab-like robe and headdress. A slightly disturbing feature of this spirit is his pair of shadow-filled, empty eye sockets. I am not really sure why he appears this way, as he will not answer my questions on this topic.

Expertise/Tasks: Frucisierre has the ability to give magical life to objects. In its simplest form, this power allows him to create powerful talismans; at full strength, it enables him to create physical egregores or golems. The technique for creating etheric egregores given in chapter 10 is not as advanced as some of the physical methods this spirit teaches; however, he will require you to keep this knowledge a secret. He will also teach you a word of power you must never write down, only memorize. This word is the key to working the rituals he teaches.

4. Forneus

Title/Type: Marquis/Goetic

Sphere: NA

Appearance: This spirit cannot assume a human form. It manifests as a large aquatic creature with gills on its neck, fins along its back, and large eyes. Forneus is not quite as menacing in appearance as the Creature from the Black Lagoon, but not exactly pleasing to look at, either.

Expertise/Tasks: Despite his unusual appearance, Forneus is a helpful spirit to work with if you are trying to master a foreign language. From giving you tips on studying to showing you how to magically speed up your learning process, Forneus can help you learn any language in a relatively short period of time. Also, if you have a document written in another language, this spirit can translate it for you as you read it out loud.

5. Marbas

Title/Type: President/Goetic

Sphere: NA

Appearance: Marbas appears as a large lion and at your command will transform into a man with the head of a lion. However strange his appearance may seem, there is nothing at all intimidating about it.

Expertise/Tasks: This friendly spirit can cure almost any disease. When you send Marbas off to cure a sick person, the afflicted will have a dream of a timid and friendly lion and will wake up feeling much better. Aside from his healing abilities, Marbas is also very knowledgeable about architecture and mechanical engineering. With his areas of expertise, Marbas is able to find and correct flaws in both humans and mechanical devices.

6. Hiepacth

Title/Type: NA

Sphere: NA

Appearance: Hiepacth appears as a dark-skinned woman with somewhat Egyptian-looking features. She has long black hair, large, penetrating eyes, and wears a black and gray cloak, which hides her body.

Expertise/Tasks: This spirit can help you find a missing person, provided you have some kind of physical link, such as an object that belonged to him or her or a signature. For some reason, Hiepacth will not work with a photograph. You must summon Hiepacth to the physical plane for her power to work effectively. When you do so, slide the chosen object into the triangle so she can travel to the missing person through the link established in this manner. When this happens, Hiepacth will vanish for a few moments, and return with news of the missing person's location. If the person you are trying to find is hiding on purpose, and Hiepacth feels he or she is justified in doing so, the form of her answer may vary from a puzzle to no answer at all. Hiepacth respects the privacy of those she searches for.

7. Naberius

Title/Type: Marquis/Goetic

Sphere: NA

Appearance: Naberius cannot assume a human form. He always appears as a black crane.

Expertise/Tasks: This spirit is very knowledgeable in several arts and sciences. If you need help studying for a test or would like a complex theory or principle explained, Naberius is an excellent spirit to work with.

8. Minoson

Title/Type: NA

Sphere: NA

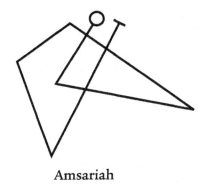

Amsariah

Michael

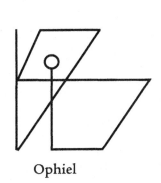

Ophiel

Hihaiah

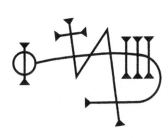

Gabriel

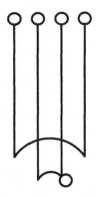

Phul

SIGIL SHEET 6

Appearance: Minoson appears as a man with a large, bald head and round blue eyes. He wears only green pants.

Expertise/Tasks: This spirit is the master of all types of statistics and can help you tabulate all sorts of data, including the odds of winning at certain games of chance. After evoking him, wearing his charged sigil when gambling will make it easier for you to win. Minoson probably won't make you very rich, but at the very least his help should guarantee you win at games of chance at a steady rate, which ensures you always come out ahead.

9. Seere

Title/Type: Prince/Goetic

Sphere: NA

Appearance: Seere appears as a well-dressed, jewel-adorned man riding a white, winged horse.

Expertise/Tasks: This spirit makes an excellent scout as well as information-gatherer. He can travel to any location on the planet and return almost instantaneously with whatever information you request he find. Seere is very helpful and seems almost pleased to go on such journeys. He is one of the friendliest non-angelic entities you can work with.

10. Nemod

Title/Type: NA

Sphere: NA

Appearance: Nemod appears as a bald old man wearing a silver robe. His face remains completely expressionless at all times, which may make you feel a little uneasy when you first communicate with him.

Expertise/Tasks: Nemod is a helpful spirit to work with if you practice evocations. As I've mentioned earlier, spirits often lie, and it is helpful to know if they are telling the truth or not. Nemod can teach you several ways to determine the sincerity of spirits and, remarkably, people as well. Before you discuss an important topic with someone, you may want to evoke Nemod to make sure you know if you're hearing the truth. The methods Nemod teaches for finding the truth may seem simple, but they are effective.

11. Purson

Title/Type: King/Goetic

Sphere: NA

Appearance: Purson appears as a man with a lion's head, carrying a snake in his right hand and riding a large, black bear.

Expertise/Tasks: This spirit is very wise and seems almost omniscient when he speaks of history and the future. He can tell you tales of how ancient people lived and give you the locations of their civilizations and treasures. If you like reading about history, you will find communicating with this spirit fascinating. Purson can also predict the immediate future with extreme accuracy, although this power seems to be limited to random events, and he can only rarely answer specific questions regarding the future.

12. Ose

Title/Type: President/Goetic

Sphere: NA

Appearance: When first evoked, Ose appears as a large, black leopard. At your command, he will take on the appearance of a man with dark features, wearing a furry, black cloak.

Expertise/Tasks: Ose can teach you the amazing art of shapechanging. Of course, mastery of this technique on the physical plane will take many years of practice, but mastering it on the astral plane can have some wonderful effects as well. For example, if you ever feel insecure, changing the shape of your astral body to that of a dragon will make aggressors feel the need to stay away, even though they can't understand why.

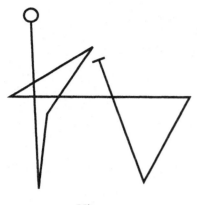

Vimarone

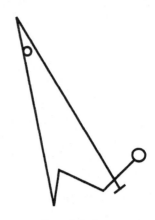

Sandalphon

Sirchade

Bael

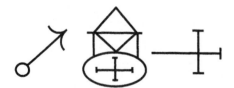

Frucisierre

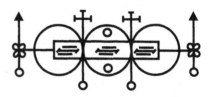

Forneus

SIGIL SHEET 7

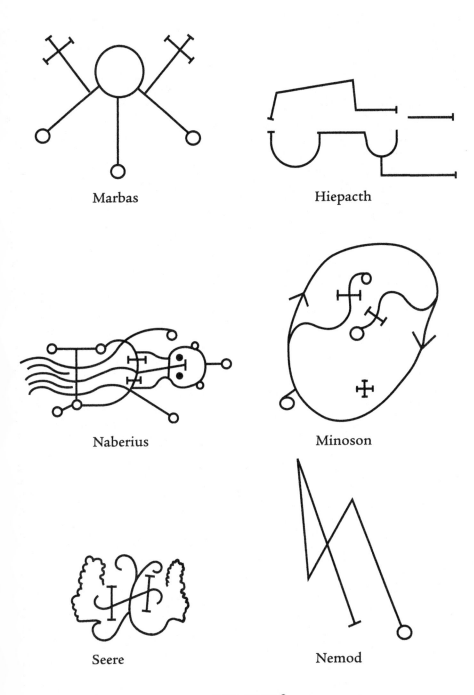

Marbas

Hiepacth

Naberius

Minoson

Seere

Nemod

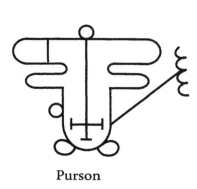

Purson

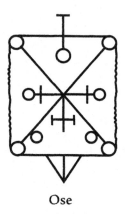

Ose

SIGIL SHEET 9

chapter ten

EGREGORE CREATION

> While sculpting, concentrate on the
> attributes and function of the being you
> are creating a representation of, as this
> step in the process is a lot more magical
> than it first seems.

O nce you have mastered both types of
evocation and have worked with several
of the spirits in chapter 9, you may want to
move on to a more advanced magical technique. Therefore, in this chapter, we'll be
dealing with the esoteric art of creating and
commanding mystical beings known as egregores; a practice that has few limitations and
almost limitless possibilities.

If you think back to chapter 1, you'll recall
the legend of the golem, in which a magically
created being became uncontrollable. This
story introduced a magical entity known as

the egregore, which is a being a magician creates to carry out some specific appointed task. Unlike the golem, however, the type of egregore
you will learn to create in this chapter is etheric and only rests in a physical shell when not active. If you would like to learn how to create physical egregores, I suggest working with the spirit Frucisierre described in
chapter 9.

By creating an etheric egregore using the technique in this chapter, and by following the instructions regarding its eventual destruction, you should have no problems similar to those Rabbi Judah Loew
faced. However, before you continue, reread the golem story to make
sure you understand the danger of being irresponsible when creating
such a being.

The Eight Steps of Creating an Egregore

The process of creating and eventually destroying an egregore consists
of eight steps: (1) deciding what task it will accomplish; (2) deciding on
its appearance; (3) naming it; (4) creating its sigil; (5) sculpting a representation of it to reside in when not active; (6) performing a ritual to
give it life for a predetermined amount of time; (7) performing a ritual
to destroy it when that time comes; and (8) disposing of the physical
representation.

The first step in bringing an egregore to life is to find a purpose for
its existence. In other words, you have to have some task or function
for the egregore to carry out. This could be anything from protecting a
certain room or area to helping a certain horse run faster in a race. Of
course, the egregore's purpose should not be an evil one, as you will
suffer the karmic consequences of its actions. Once you come up with
a suitable task for your egregore to accomplish, try to word it as a command, similar to the types of commands you give spirits in evocations.
Make sure your statement of its purpose is clear and concise, as it is the
instruction you will give the egregore when it is created.

The next step you have to follow is relatively easy. Simply come up
with an appearance you want your egregore to have. Use some common sense here. Don't make it look so ferocious that you can barely
stand to look at it when you evoke it. Try to come up with a simple

appearance that matches the being's function. For example, if the egregore's function is protection, give it a weapon of some kind, while if its purpose is urging on a horse, you may want to give it a crop (an egregore will not actually use these types of implements, rather they will act as symbols of its function). You can create egregores that do not look human, but try to do this for a symbolic reason. For example, if you want help becoming a better swimmer, create an egregore that has webbed hands and feet.

Once you know the purpose and appearance of the egregore you are trying to create, the next step is to come up with a name for it. The particular name you choose doesn't really matter, as long as it is not a name that means something contradictory to the egregore's purpose. For example, if you see a movie where the villain is named Dagon, do not give this name to an egregore whose purpose is to help alleviate someone's suffering. If the name you choose already means something negative to you, then you might end up feeding some of that negativity into your creation. It's better to either make up a name or find one that reflects the qualities of your egregore. For example, if you're an admirer of medieval literature, you might want to use the name Roland for an egregore that is supposed to help you increase your level of bravery.

After you decide upon your creation's name, you are ready to create its sigil. To do so, you should use the Rose Cross Lamen method given in chapter 6 (when creating egregore sigils, use the Rose Cross in Roman characters, unless you are knowledgeable in Hebrew and wish to spell your egregore's name in this alphabet). For now, you only have to design the sigil on a piece of paper, but during the ritual birth of your egregore, you will be required to add this sigil to your physical representation of the being. Make sure this sigil is correct in accordance with the spelling you choose for your egregore's name.

The last step of your egregore's creation before the ritual of birth given below is performed, is to create a physical representation of the being. To do this, you will need nonhardening modeling clay. You can find this in almost any art supply or variety store. As you will soon see, it is very important to use a clay that will not harden on its own. If you have decided upon a particular color that you predominantly want your egregore to be, then you can use this color clay. If you haven't given much thought to color or you can't find the appropriate color, you can use either brown, off-white, or gray clay.

Once you find your clay, spend some time meditating upon the image of the egregore you have built up in your mind. Sketch it if it helps. When you feel you are certain of its appearance, use the clay to try to sculpt a model of it. It should be about eight or nine inches in height and about three or four inches wide, as this is an easy size figure to work with. While sculpting, try to concentrate on the attributes and function of the being you are creating a representation of, as this step in the process is a lot more magical than it first seems.

Here are some things to keep in mind when sculpting. If the egregore is of a nonhuman shape, try to sculpt a figure that shows some of your egregore's most outstanding features (for example, wings or long ears). When dealing with a basic human-shaped being, try to add any features that make the egregore unique (for example, tools or implements it may carry). Your figure does not have to look as if a master or even apprentice sculptor created it. The sculpture only has to bear a strong enough resemblance to the egregore so your imagination can easily fill in the missing details when you look at it.

With your sculpture prepared, you are ready to perform the ritual that will give life to your egregore. Set up your temple in the way described below, which is similar to the preparation for an evocation to the physical plane, with a few changes. First, put your Triangle on the floor within the east quarter of your circle. You will have to slide it out of the circle during the ritual. Place your censer within the Triangle and light the charcoal. Prepare a mixture of gum mastic and frankincense for your censer, as this is a heavy-bodied incense that works well for manifestation, but do not add it until you are told to do so in the ritual. Set your lamp outside the circle and behind you so it dimly illuminates the room. Do not use a filter.

Your altar should be set up in the following fashion. On your Tablet of Union, place the sculpture you made and a piece of paper with your egregore's name and sigil written upon it in black ink. Both of these should be covered with one piece of white silk. Also have a toothpick, a lit candle, a dish of salt, a bowl of water, and a lit censer on your altar. Like the censer used in the Triangle, do not add incense to this censer until the time comes to do so in the ritual. Finally, place your regular magical implements on the altar as well.

Copy the necessary orations from the ritual below into your book and have it handy to read from. There is no need to memorize any of them. With this last preparation completed, you are ready to begin the ritual creation of your egregore. In this ritual, you will have to assign the egregore an amount of time to complete its task. I like to allow these beings one month to carry out their tasks before I terminate their magical existence. If you would like to work with one egregore for a longer period of time, make sure you summon it at least once a month to command it to give you a report of its success. In either case, make sure you have an exact day for the termination of your egregore's existence decided upon before you start the following ritual.

Ritual for Egregore Creation

1. Set up your temple and altar in the fashion described above. Put on your robe and ring if you have them.

2. Perform the Opening by Watchtower up to and including step 17 (see chapter 3).

3. Uncover the sculpture and paper on your Tablet of Union. Lift up the sculpture and say the following: *O Creature of Clay, before thou canst be given life, thou must be made pure by the elements.*

4. Pass the sculpture over the lit candle and say, *I purify thee with Fire.* Now, pick up your Fire Wand and wave it over the sculpture three times.

5. Replace your Wand and, using your fingers, sprinkle some water from your dish onto the sculpture, saying,

I purify thee with Water. Pick up your Water Cup and wave it over the sculpture three times.

6. Replace your Cup. Add incense to your altar censer and pass the sculpture through the smoke, saying, *I purify thee with Air.* Wave your Air Dagger over the sculpture three times.

7. Replace your Dagger. Touch the base of your sculpture to the dish of salt and say, *I purify thee with Earth.* Wave your Earth Pentacle over the sculpture three times.

8. Replace your Pentacle. Hold up the sculpture and say the following: *O Creature of Clay, purified by the elements and ready to receive life, I now hold you up to the Light whose service you shall be committed to.* Visualize white light descending upon your sculpture, then look up to the heavens and add: *O Glorious One, may this creature be granted life with your permission.*

9. Pick up your toothpick and inscribe your egregore's name onto its back while saying, *I name thee, N., and from this day until (give time) on (give date) you shall be known by this name, and shall be given life to perform the task which I will assign to you.*

10. Below the egregore's name, use the toothpick to inscribe the sigil, saying, *With this sigil, I shall most easily be able to contact you, N., but when its lines are no more, so too shall your life and purpose be no more, for this is the will of the One True God.*

11. Pick up the sculpture and perform the Middle Pillar Ritual. Channel some of the energy raised into the sculpture with the knowledge that this being is linked to you until the time of its ritual death.

12. Walk clockwise to the Triangle in the east of your circle, carrying your Magic Sword and the sculpture. Stand the sculpture behind the censer in the Triangle. Add a good amount of incense to the censer and, using your Sword, slide the Triangle out of your circle to a distance of two feet or as far as the length of your Sword allows, if it is shorter. *Note:* If you do not have a wooden Triangle, I suggest you place it upon a piece of wood for this ritual and push the wood with your sword, as a thin cardboard Triangle might bend when you try to slide it out of the circle.

13. Return to your position behind the altar, facing east. Replace your Magic Sword. Pick up the paper sigil of the egregore and your book in your left hand and your Magic Wand in your right. Return to the east of your circle, moving clockwise.

14. Try to visualize the features of your egregore on the sculpture. Point your Wand at the Triangle and give the following oration: *By the power of the Most High, and through the influence of the elements that have purified your vessel, I now call you into being, N. Know that your purpose in existing is to serve me by performing the task of* (state the task). *I hereby command you to complete these duties by* (give the date and time), *and to reside within this vessel of clay whenever you are not actively pursuing these duties. At no time shall you delay your justified return to this vessel for an instant, and in no way shall you cause harm to befall anyone. This I command with the authority of the One True God, Whose Path of Light you must follow.*

15. Gaze at the sigil of the egregore in your hand and try to visualize it hovering in the smoke before the sculpture. Recite the following conjuration: *Having given thee magical life, I command you to appear before me within this Triangle in your newly granted form. Rise from your physical clay shell, and make your understanding of your purpose*

known to me. Appear, N., for it is the Lord Whose Light has given you life Who commands you now.

16. Enter your astral body of light and once again visualize the sigil of the entity floating before the sculpture. Call out to the egregore, repeating its name as you see it emerge from the sculpture. It will look like a mist rising from the clay figure at first, but it should gradually begin to assume the form you created for it. As it starts to take shape, you can begin to whisper its name and slowly open your eyes, as in any evocation to the physical plane. The egregore will be standing before you when you open your eyes.

17. Communicate with the egregore and make certain it understands what its function will be.

18. When you are ready to send the egregore off, say the following: *Go in peace, N., and begin your labors. Remember to reside within this vessel of clay when idle, and be certain to never harm another. Be ready to appear quickly before me when called, and to give an accurate report of your progress.*

19. Perform step 18 of the Watchtower Ritual.

20. Perform the Closing by Watchtower.

21. Wrap the sculpture in the black piece of silk, along with the paper the name and sigil are drawn upon. Put this in a place where no one will touch it.

When you wish to contact your egregore, perform a normal evocation, using the method given in step 16, as the egregore does not reside in another sphere. The sculpture should always be placed in the Triangle for these evocations, and the paper sigil used in the creation should be the one gazed at each time. Do not recopy the sigil.

When the time comes for the destruction of your egregore, set up your temple in the same way you set it up for the creation ritual. This time the Triangle will not be needed, however, as the egregore should never be summoned to visible appearance on the day of its termination.

Ritual for Egregore Destruction

1. Prepare your temple as described above. Wear your robe and ring.

2. Perform the Opening by Watchtower up to and including step 17.

3. Unwrap the sculpture and paper on your Tablet of Union. Lift the sculpture, look to the heavens, and say, *Many thanks, O Glorious One, for allowing this servant, N., to aid me in the workings of the Path of Light.*

4. Lower the sculpture and say over it, *Your duty completed, and your predestined time of death reached, I now return your essence and being to the Universe, for the source that grants life also takes it away. This is a Divine Mystery, and the Authority of the Lord cannot be disputed.*

5. Wet your finger in the bowl of water, and smear the sigil inscribed on the back of the figure, while saying, *As these lines are washed away by Water, so too does thine existence cease, N., for this is the will of the One True God. This sigil shall no longer summon thee.*

6. Smear the name inscribed on the back of the figure, while saying, *You are no more, Creature of Clay, and this name shall no longer summon thee.*

7. Try to feel energy leave the sculpture. Rest it on the Tablet of Union and pick up the piece of paper bearing the name and sigil of the egregore. Light this with the flame of the candle and drop it into the censer while *saying, The energy of this magical being is now returned to the Source of All.*

8. Lift the sculpture and pass it over the flame of the candle, saying, *I cleanse this clay with Fire.*

9. Using your fingers, sprinkle some water over the sculpture and say, *I cleanse this clay with Water.*

10. Add incense to the censer and pass the sculpture through the rising smoke, saying, *I cleanse this clay with Air.*

11. Place the figure on the dish of salt and push down on it, crushing it, while saying, *I cleanse this now-formless clay with Earth, and by doing so, return it to its element of origin.*

12. Perform step 18 of the Watchtower Ritual.

13. Perform the Closing by Watchtower.

14. Dispose of the clay as described below.

The lump of clay you are left with at the close of this ritual must be properly returned to the earth. To do so, find a secluded area and dig a deep hole. Lay this clay in the hole and sprinkle the ashes of the paper from your censer on top of it. Then fill in the hole, making sure to leave the area as it was when you found it. With this last step completed, your egregore is no more.

Let me take a moment to explain the importance of not letting an egregore live on after its usefulness and time have run out. These beings have a desire to survive that is of a dangerous intensity. If you were to allow an egregore to live for too long, its strength would increase until controlling it would become impossible. Eventually, a renegade egregore of this type would roam the astral plane, free to do anything it desired. Remember, egregores have no "place" in our universe, and do not fit into the existing order and hierarchy of entities.

The existence of an egregore after it no longer has a purpose to serve would be a bizarre one. Once it became aware of its inability to assimilate into the order of magical beings, an egregore would begin to display highly erratic behavior. You see, when I mention the order of entities, I am not only referring to a group, but to a state of being as well. There is no order in the existence of an egregore that no longer has a purpose, only chaos. Even an egregore created to bring peace to a relationship could become a mischievous entity worthy of its own horror movie if it were allowed to exist without having to perform its function any longer.

In chapter 1 we dealt with egregores that are created by groups or lodges to exist within specific magical currents. These beings are not dangerous in the manner explained above, and for this reason, survive for decades and often centuries, helping magicians who work within their current. Do not confuse these beings with an egregore that is created by a magician for a specific purpose. This latter type of egregore, which is the subject of this chapter, can only serve the magician who creates it. Unfortunately, the damage it can cause will most likely also center around this magician. To make things easier and avoid potential disaster, adhere to the following warning/piece of advice: Do not let one of these beings exist for too long! As I've said earlier, a month is perfect. At most, never extend the life of an egregore beyond three or four months.

The process for the creation and destruction of egregores explained in this chapter requires more effort to perform successfully and takes longer to complete than does an evocation. It is, however, worth the extra time and work to utilize these beings, as you can command them to perform any type of task you desire. This is very helpful because you may not always be able to find a preexisting spirit that can do exactly what you would want it to. If you heed the above warnings and work this type of magic responsibly, you will find it to be extremely rewarding.

CONCLUSION

After the magician read the license to depart, he watched as the figure in the Triangle began to dissipate. When Phalegh fully disappeared from sight, the magician did the necessary closing ritual and left the room to change into his new suit. Fifteen minutes after the altar lamp was extinguished, the magician was at his front door.

Outside, with the cold winter air returning him to normal consciousness, the magician walked a few blocks north to the train station and climbed the concrete stairs. The morning crowd had already gathered, waiting for the train that was now five minutes late for some reason. As the steel vehicle came into sight and approached the platform, the magician thought of the interview he was on his way to, and how it could lead to his first job in a seemingly heartless city. Over the past few months, he had often wondered what challenges this new time in his life would bring—thoughts that had often resulted in a fair share of sleepless nights and anxiety.

But today would be different. With Phalegh's sigil in his pocket and a new sense of courage filling him, he entered the train . . .

The story at the beginning and end of this book is only an example of how evocation can improve one's life. Whether you need a little help facing one of life's challenges, as the magician above, or you need to work a minor miracle for someone in a hospital miles away, I hope you find in the art of magical evocation a powerful tool for change.

Good luck in all your mystical endeavors and may your evocations bring you much happiness.

BIBLIOGRAPHY

Achad, Frater. *Crystal Vision Through Crystal Gazing.* Chicago: Yogi Publication Society, 1923.

Bardon, Franz. *Frabato the Magician.* Wuppertal, West Germany: Dieter Ruggeberg Verlag, 1982.

———. *Initiation into Hermetics.* Wuppertal, West Germany: Dieter Ruggeberg Verlag, 1987.

———. *The Key to the True Quabbalah.* Wuppertal, West Germany: Dieter Ruggeberg Verlag, 1986.

———. *The Practice of Magical Evocation.* Wuppertal, West Germany: Dieter Ruggeberg Verlag, 1984.

Barrett, Francis. *The Magus: A Complete System of Occult Philosophy.* New York: Citadel Press, 1989.

Buckland, Raymond. *Buckland's Complete Book of Witchcraft.* St. Paul, Minn.: Llewellyn Publications, 1990.

———. *Doors to Other Worlds.* St. Paul, Minn.: Llewellyn Publications, 1993.

Cicero, Chic and Sandra Tabitha. *Secrets of a Golden Dawn Temple: The Alchemy and Crafting of Magickal Implements.* St. Paul, Minn.: Llewellyn Publications, 1992.

Crowley, Aleister. *Goetia.* New York: Magickal Childe Publishing, 1989.

———. *777 and Other Qabalistic Writings of Aleister Crowley.* York Beach, ME: Samuel Weiser, 1991.

Denning, Melita, and Osborne Phillips. *Planetary Magick.* St. Paul, Minn.: Llewellyn Publications, 1989.

Hay, George, ed. *The Necronomicon: The Book of Dead Names.* London: Skoob Books Publishing, 1992.

Honorius. *The Sworn Book of Honorius the Magician.* New Jersey: Heptangle Books, 1977.

Kraig, Donald Michael. *Modern Magick: Eleven Lessons in the High Magickal Arts.* St. Paul, Minn.: Llewellyn Publications, 1989.

Levi, Eliphas. *Transcendental Magic: Its Doctrine and Ritual.* New York: Samuel Weiser, 1974.

Malchus, Marius. *The Secret Grimoire of Turiel.* London: Aquarian Press, 1971.

Mathers, S. Liddell MacGregor, ed. *The Book of the Sacred Magic of Abra-Melin the Mage.* New York: Causeway Books, 1974.

———. *The Grimoire of Armadel.* York Beach, Maine: Samuel Weiser, 1980.

———. *The Key of Solomon the King.* York Beach, Maine: Samuel Weiser, 1990.

McLean, Adam, ed. *A Treatise on Angel Magic.* Grand Rapids, Mich.: Phanes Press, 1990.

Regardie, Israel. *The Golden Dawn.* St. Paul, Minn.: Llewellyn Publications, 1989.

Schueler, Gerald, and Betty. *An Advanced Guide to Enochian Magick.* St. Paul, Minn: Llewellyn Publications, 1988.

———. *Enochian Magic: A Practical Manual.* St. Paul, Minn: Llewellyn Publications, 1990.

Shaw, Idries. *The Secret Lore of Magic.* New York: Citadel Press, 1958.

Simon, ed. *The Necronomicon.* New York: Avon Books, 1980.

Three Initiates. *The Kybalion.* Chicago: Yogi Publication Society, 1940.

Trismegistus, Hermes Mercurius. *Divine Pymander.* Boston: Rosicrucian Publishing Company, 1871.

Turner, Robert, ed. *The Arbatel of Magic.* New Jersey: Heptangle Books, n.d.

Waite, Arthur Edward. *The Book of Ceremonial Magic.* New York: Citadel Press, 1990.

Zalewski, Pat. *The Kabbalah of the Golden Dawn.* St. Paul, Minn.: Llewellyn Publications, 1993.

INDEX

CPSIA information can be obtained at www.ICGtesting.com
Printed in the USA
BVOW06s1306230616

453119BV00003B/3/P